Introduction to probability theory

KIYOSI ITÔ

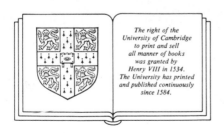

The right of the
University of Cambridge
to print and sell
all manner of books
was granted by
Henry VIII in 1534.
The University has printed
and published continuously
since 1584.

CAMBRIDGE UNIVERSITY PRESS
Cumbridge
London   New York   New Rochelle
Melbourne   Sydney

Published by the Press Syndicate of the University of Cambridge
The Pitt Building, Trumpington Street, Cambridge CB2 1RP
32 East 57th Street, New York, NY 10022, USA
296 Beaconsfield Parade, Middle Park, Melbourne 3206, Australia

Introduction to Probability Theory by Kiyosi Itô

Originally published in Japanese by Iwanami–Shoten, Publishers,
Tokyo, 1978

Library of Congress Cataloging in Publication Data
Itô, Kiyosi, 1915–
Introduction to probability theory.
Translation of: Kakuritsuron. 1-4-shō.
Includes index.
1. Probabilities.
QA273.I822513 1984   519.2   83-23187
ISBN 0 521 26418 9 hardcover
ISBN 0 521 26960 1 paperback

Contents

Preface   v

Notation and abbreviations   vii

1 Finite trials   1
1.1 Probability spaces   1
1.2 Real random variables and random vectors   3
1.3 Mixing, direct composition, and tree composition   13
1.4 Conditional probabilities   24
1.5 Independence   27
1.6 Independent random variables   32
1.7 The law of large numbers   35

2 Probability measures   38
2.1 General trials and probability measures   38
2.2 The extension theorem of probability measures   46
2.3 Direct products of probability measures   53
2.4 Standard probability spaces   60
2.5 One-dimensional distributions   67
2.6 Characteristic functions   80
2.7 The weak topology in the distributions   97
2.8 d-Dimensional distributions   102
2.9 Infinite-dimensional distributions   105

3 Fundamental concepts in probability theory   110
3.1 Separable perfect probability measures   110
3.2 Events and random variables   114
3.3 Decompositions and σ-algebras   123
3.4 Independence   129
3.5 Conditional probability measures   136
3.6 Properties of conditional probability measures   145
3.7 Real random variables   148
3.8 Conditional mean operators   157

iii

4 Sums of independent random variables 165
4.1 General remarks 165
4.2 Convergent series of independent random variables 170
4.3 Central values and dispersions 175
4.4 Divergent series of independent random variables 184
4.5 Strong law of large numbers 186
4.6 Central limit theorems 191
4.7 The law of iterated logarithms 199
4.8 Gauss's theory of errors 206
4.9 Poisson's law of rare events 209

Index 212

Preface

Because a point in space can be represented by a triple of real numbers, all geometric properties of spatial figures can be expressed in terms of real numbers. Hence one can theoretically understand geometry solely through analysis. But a true appreciation of geometry requires not only analytical technique but also intuition of geometric objects. The same holds for probability theory. The modern theory of probability is formulated in terms of measures and integrals and so is part of modern analysis from the logical viewpoint. But to really enjoy probability theory, one should grasp the orientation of development of the theory with intuitive insight into random phenomena. The purpose of this book is to explain basic probabilistic concepts rigorously as well as intuitively.

In Chapter 1 we restrict ourselves to trials with a finite number of outcomes. The concepts discussed here are those of elementary probability theory but are dealt with from the advanced standpoint. We hope that the reader appreciates how random phenomena are discussed mathematically without being annoyed with measure-theoretic complications.

In the subsequent chapters we expect the reader to be more or less familiar with basic facts in measure theory.

In Chapter 2 we discuss the properties of those probability measures that appear in this book.

In Chapter 3 we explain the fundamental concepts in probability theory such as events, random variables, independence, conditioning, and so on. We formulate these concepts on a perfect separable complete probability space. The additional conditions "perfectness" and "separability" are imposed to construct the theory in a more natural way. The reader will see that such conditions are satisfied in all problems appearing in applications.

In the standard textbook conditional probability is defined with respect to σ-algebras of subsets of the sample space (Doob's definition). Here we first define it with respect to decompositions of the sample space

(Kolmogorov's definition) to make it easier for the reader to understand its intuitive meaning and then explain Doob's definition and the relation between these two definitions.

In Chapter 4 we discuss the properties of infinite sums of independent real random variables.

At the end of each section several problems are presented with hints for solution to help the reader understand the material in the section.

This book is the English version of the first four chapters of my book *Probability Theory* (in Japanese, Iwanami-Shoten, Tokyo, 1978), based on my course of probability theory at Kyoto University (Japan), at Aarhus University (Denmark), and at Cornell University (U.S.A.). I am grateful to my colleagues and students for their valuable comments. Also I thank David Tranah at Cambridge University Press for his kind cooperation and Mrs. H. Shinohara for her painstaking job of typing.

Kiyosi Itô
Tokyo

Notation and abbreviations

Symbols and terms marked ($*$) may not be in universal usage and so should be given special attention.

General

($*$) topological space $= T_1$-space
($*$) increasing $=$ nondecreasing
($*$) positive-definite $=$ non$-$negative-definite
($*$) countable $=$ at most countable

Set theory

$\emptyset =$ the empty set
($*$) $2^A =$ the power set of A
($*$) $\#A =$ the cardinal number of A
$A^c =$ the complement of A
$\cup =$ union
($*$) $+, \Sigma =$ disjoint union
($*$) $- =$ proper difference
($*$) $\Delta =$ symmetric difference
$\times, \Pi =$ Cartesian product
($*$) $\pi_k =$ projection to the kth component space
$f = A \rightarrow B, \; x \rightarrow f(x) =$ the map f from A into B carrying x to $f(x)$
$\mathscr{D}(f) =$ the domain of definition of f
$f(C) =$ the image of C under f
$f^{-1}(D) =$ the inverse image of D under f

Analysis

$\mathbf{N} =$ the natural numbers
$\mathbf{Q} =$ the rational numbers

R = the real numbers

C = the complex numbers

$[a, b) = \{x \in \mathbf{R} | a \le x < b\}$

$(a, b), (a, b], [a, b]$: defined similarly

$\mathbf{R}^n$ = the n-space

(∗) $\mathbf{R}^\infty$ = the sequence space

(∗) $\binom{n}{r} = {}_nC_r = n!/r!(n-r)!$

$[x] = \max\{n \in N | n \le x\}$ (Gauss bracket)

$x \vee y = \max(x, y)$

$x \wedge y = \min(x, y)$

$x^+ = x \vee 0$

$x^- = (-x) \vee 0$

Re = real part

Im = imaginary part

ess. sup. = essential supremum

sup = supremum (least upper bound)

inf = infimum (greatest lower bound)

lim sup = limit superior (upper limit)

lim inf = limit inferior (lower limit)

(∗) $f(a+) = f(a+0) = \lim_{\varepsilon \downarrow 0} f(a+\delta)$

(∗) $f(a-) = f(a-0) = \lim_{\varepsilon \downarrow 0} f(a-\varepsilon)$

(∗) $1_A = \chi_A$ = the indicator (or characteristic function) of A

supp = support (of a function)

(∗) $f(x)|_{x=a}$ = the value of $f(x)$ evaluated at $x = a$

$f|_A$ = the restriction of f to A

$\mathscr{F}$ = Fourier transform

a.e. = almost everywhere

$\| \ \|_p = p$th-order norm

$L^p(X, m)$ = the L^p-space over (X, m)

(∗) $mf^{-1} = fm$ = the image measure of m by f

(∗) f is measurable $\mathscr{B}_1/\mathscr{B}_2 \Leftrightarrow f^{-1}(\mathscr{B}_2) \subset \mathscr{B}_1$

Probability theory

(Ω, P) = the base probability space

$\mathscr{D}(P)$ = the domain of definition of P = the P-measurable sets

a.s. = almost surely

(∗) i.o. = infinitely often

(∗) f.e. = with a finite number of exceptions

i.p. = in probability

$A, B, C, \ldots$ = classes of subsets of Ω, i.e., subsets of 2^{Ω}

$\mathbf{2} = \{ A \in \mathscr{D}(P) : P(A) = 0 \text{ or } 1 \}$ = the trivial σ-algebra

(∗) $\delta[A]$ = the Dynkin class generated by A

(∗) $\sigma[A]$ = the σ-algebra generated by A

$\mathscr{B}(S)$ = the Borel system (topological σ-algebra) on S

$\bigwedge_{\alpha} \mathscr{B}_{\alpha}$ = the greatest σ-algebra that is included in every

$$\mathscr{B}_{\alpha} = \bigcap_{\alpha} B_{\alpha}$$

$\bigvee_{\alpha} B_{\alpha}$ = the least σ-algebra that includes every $\mathscr{B}_{\alpha} = \sigma[\bigcup_{\alpha} \mathscr{B}_{\alpha}]$

$\Delta, \Delta', \ldots$ = decompositions of Ω

(∗) $\Delta_{\mathscr{A}}$ = the decomposition generated by A

(∗) $X_{\Delta}(\omega)$ = the element of Δ that contains ω

(∗) $\mathscr{B}_{\Delta}$ = the σ-algebra generated by Δ

$\Delta \succ \Delta' = \Delta$ is finer than Δ'

$\bigvee_{n} \Delta_{n}$ = the least upper bound of $\{ \Delta_{n} \}$

$\bigwedge_{n} \Delta_{n}$ = the greatest lower bound of $\{ \Delta_{n} \}$

$\lambda, \mu, \nu, \ldots$ = n-dimensional distributions $(n = 1, 2, \ldots, \infty)$

$\mathscr{P}^{n}$ = the n-dimensional distributions

$\mu * \nu$ = the convolution of μ and ν

$\check{\mu}$ = the reflection of μ

$\tilde{\mu} = \check{\mu} * \mu$

C_{μ} = the continuity points of μ

D_{μ} = the discontinuity points of μ

$M^{p}(\mu)$ = the pth-order moment of μ

$|M|^{p}(\mu)$ = the pth-order absolute moment of μ

$M(\mu) = M^{1}(\mu)$ = the mean value of μ

$V(\mu)$ = the variance of μ

$\gamma(\mu)$ = the central value of μ

$\delta(\mu)$ = the dispersion of μ

$\varphi_{\mu} = \mathscr{F}\mu$ = the characteristic function of μ

$N_{m, v}$ = Gauss distribution

$C_{m, c}$ = Cauchy distribution

p_{λ} = Poisson distribution

δ = delta distribution

$X, Y, Z, \ldots$ = random variables

(∗) $P^{X} = PX^{-1}$ = the probability distribution of X

$E(X)$ = the expectation of $X = M(P^{X})$

$V(X)$ = the variance of $X = V(P^{X})$

$\sigma(X)$ = the standard deviation of $X = \sigma(P^{X})$

(∗) $V(X, Y)$ = the covariance of X and Y

$R(X, Y)$ – the correlation coefficient between X and Y

($*$) $\gamma(X) =$ the central value of $X = \gamma(P^X)$

($*$) $\delta(X) =$ the dispersion of $x = \delta(P^X)$

$\varphi_X =$ the characteristic function of $X = \mathscr{F}P^X$

$M_p(X) =$ the pth-order moment of $X = M_p(P^X)$

$|M|_p(X) =$ the pth order absolute moment of $X = |M|_p(P^X)$

$\Delta_X =$ the decomposition generated by X

$\sigma[X] =$ the σ-algebra generated by X

($*$) $\bar{\sigma}[X] = B_{\Delta_X} = X^{-1}(\mathscr{D}(P^X))$ $(\sigma[X] \subset \bar{\sigma}[X] = \subset \sigma[X] \vee \mathbf{2})$

($*$) $P_{\Delta_X}(P_{\mathscr{B}}) =$ conditional probability measure under $\Delta(\mathscr{B})$

($*$) $P_\Delta(P_{\mathscr{B}}^X) =$ conditional probability distribution under $\Delta(\mathscr{B})$

($*$) $E_\Delta(E_{\mathscr{B}}) =$ conditional expectation under $\Delta(\mathscr{B})$

($*$) $L^0(\Omega, P) =$ the real random variables on (Ω, P)

($*$) $\|X\|_0 = E(|X| \wedge 1)$

($*$) $L_{\mathscr{B}}^0(\Omega, P) =$ the $\mathscr{B}$-measurable random variables on (Ω, P)

($*$) $L_{\mathscr{B}}^p(\Omega, P) = \{X \in L_{\mathscr{B}}^0(\Omega, P) | \|X\|_p < \infty\}$

1

Finite trials

In probability theory an experiment, an observation, or a survey is called a *trial*. A trial is called a *finite trial* if it has a finite number of outcomes. In this chapter we will explore elementary concepts in probability theory, limiting ourselves to finite trials. For this purpose, knowledge of elementary set theory is sufficient. However, once the reader understands the fundamental concepts of probability theory through observation of finite trials, he or she can easily proceed to *infinite trials*–the subject of modern probability theory–using the results in modern analysis.

1.1   Probability spaces

In observing the number obtained when throwing a die, we have six outcomes: $1, 2, 3, 4, 5, 6$. Each outcome is called a *sample point*, and the set of all sample points is called the *sample space* for the trial. The sample points and the sample space are considered in any trial. A trial is called a finite trial or an infinite trial according to whether the sample space is a finite set or an infinite set. In this chapter we observe only finite trials, which we call simply trials.

Let T be a trial and Ω its sample space. For any set $A \subset \Omega$ we say that A *occurs* instead of saying that a sample point in A is realized in the trial. In this sense a subset A of Ω is called an *event A*.

The complement A^c of A occurs if and only if A does not occur. Hence A^c is called the *complementary event* of A. The union of A and B, $A \cup B$, occurs if and only if at least one of A and B occurs. Hence $A \cup B$ is called the *sum event* of A and B. The intersection of A and B, $A \cap B$, occurs if and only if both A and B occur. Hence $A \cap B$ is called the *intersection event* of A and B. The difference of A from B, $A \setminus B$ occurs if and only if A occurs and B does not. Hence $A \setminus B$ is called the *difference event* of A and B. The inclusion relation $A \subset B$ means that B occurs whenever A does. In this case the difference $B \setminus A$ is called the *proper difference*, written $B - A$. When we use the notation $B - A$, we understand that $A \subset B$ is implicitly assumed. The events A and B are called *exclusive* if and only if the sets A and B are exclusive; that is, $A \cap B = \varnothing$. In this case $A \cup B$ is called the *direct sum* of A and B, written $A + B$. When we use the notation $A + B$, we understand that $A \cap B = \varnothing$ is assumed implicitly. Similarly for the sum $A_1 \cup A_2 \cup \cdots \cup A_n$ (or $\bigcup_{i=1}^{n} A_i$) and the direct sum $A_1 + A_2 + \cdots + A_n$ (or $\sum_{i=1}^{n} A_i$).

Let T be a trial and Ω its sample space. For any set $A \subset \Omega$ we denote by $P(A)$ the *probability* that the event A occurs. By the intuitive meaning of probability it is natural to impose the following on $P(A)$.

(p.1) $P(A) \geqq 0$,

(p.2) $P(A + B) = P(A) + P(B)$,   (additivity)

(p.3) $P(\Omega) = 1$.

In general a set function $P(A)$, $A \subset \Omega$, having these properties is called a *probability measure* on Ω, and a set Ω endowed with a probability measure P on Ω is called a *probability space* (Ω, P). Hence (Ω, P) is called the probability space for a trial T, if Ω is the sample space of T and $P(A)$ is the probability of occurrence of $A \subset \Omega$. Conversely, for any given probability space (Ω, P) we can consider a trial T of drawing a sample point from Ω with probability $P(A)$ for every $A \subset \Omega$, so that the probability space for T turns out to be (Ω, P).

Since the probabilistic aspect of a trial is completely determined by its probability space, any trials having the same probability space are identified in probability theory. Suppose that T_1 is the trial of observing the number obtained when throwing a die and T_2 is the trial of observing the number on a card drawn at random from a box with six cards numbered 1, 2, 3, 4, 5, and 6. Then T_1 and T_2 have the same probability space (Ω, P), where

$$\Omega = \{1, 2, 3, 4, 5, 6\}, \qquad P(A) = \#A/6 \ (\#: \text{the number of points})$$

Hence T_1 and T_2 are identical with each other. In view of this observation, a trial with the probability space (Ω, P) is called simply a trial (Ω, P).

Theorem 1. If P is a probability measure on Ω, then we have the following:

(i) $P(\sum_{i=1}^{n} A_i) = \sum_{i=1}^{n} P(A_i)$,

(ii) $P(B - A) = P(B) - P(A)$,

(iii) $P(A^c) = 1 - P(A)$,

(iv) $P(A \cup B) = P(A) + P(B) - P(A \cap B)$,
(v) $P(A) = \Sigma_{\omega \in A} P\{\omega\}$ (Hence P is determined if $P\{\omega\}$ is determined for every $\omega \in \Omega$.)

Proof:
  (i) Derive this from (p.2) by induction on n.
  (ii) Apply (p.2) to $B = (B - A) + A$. Remember that $A \subset B$ is assumed implicitly when we use the notation $B - A$.
  (iii) Set $B = \Omega$ in (ii).
  (iv) Setting $C = A \cap B$, $A_1 = A - C$, and $B_1 = B - C$, we have

$$A = A_1 + C, \qquad B = B_1 + C, \qquad A \cup B = A_1 + B_1 + C,$$

  which implies (iv) by (i).
  (v) Apply (i) to $A = \Sigma_{a \in A}\{a\}$. ∎

We have mentioned that an event is represented by a set. We also represent an event by a condition. Let $\alpha(\omega)$ be a condition concerning a generic sample point ω. We say that α occurs if a sample point ω satisfying $\alpha(\omega)$ is realized as an outcome of the trial in consideration. In view of this, a condition $\alpha(\omega)$ is often called an event $\alpha(\omega)$. If we set $A = \{\omega | \alpha(\omega)\}$, then occurrence of α is equivalent to that of A. Hence the probability of occurrence of α, written $P(\alpha)$, is equal to $P\{\omega | \alpha(\omega)\}$.

The negation of α, written $\alpha^\neg$, is the complementary event of α, because $\{\omega | \alpha^\neg(\omega)\} = \{\omega | \alpha(\omega)\}^c$. The condition "$\alpha$ or β", written $\alpha \vee \beta$, is the sum event of α and β, because

$$\{\omega | \alpha(\omega) \vee \beta(\omega)\} = \{\omega | \alpha(\omega)\} \cup \{\omega | \beta(\omega)\}.$$

Similarly the condition "α and β," written $\alpha \wedge \beta$, is the intersection event of α and β.

Exercise 1. Prove the following *inclusion–exclusion formulas*:
  (i) $P(\bigcup_{i=1}^n A_i) = \Sigma_{k=1}^n (-1)^{k-1} \Sigma_{i_1 < i_2 < \cdots < i_k} P(\bigcap_{\kappa=1}^k A_{i_\kappa})$,
  (ii) The formula obtained by switching $\cup$ and $\cap$.
[*Hint*: Derive these formulas from Theorem 1(iv), using induction on n.]

1.2   Real random variables and random vectors

Let (Ω, P) be the probability space for a trial T. A real-valued function $X(\omega)$ defined on Ω is called a *real random variable* on (Ω, P), which intuitively means a quantity varying in accordance with the result of T. As we mentioned in the last section, the trial of observing the number obtained when throwing a die is represented by the probability space

$$\Omega = \{1,2,3,4,5,6\}, \qquad P(A) = \#A/6.$$

If we are to get a prize of an amount double the number obtained, the prize is a random variable $X(\omega)$ defined by $X(\omega) = 2\omega$.

The trial of observing the numbers obtained when throwing a die twice is represented by the probability space

$$\Omega = \{(i, j) | i, j = 1,2,3,4,5,6\} = \{1,2,3,4,5,6\}^2,$$

$$P(A) = \#A/36.$$

If we denote by X_1, X_2, and X the number obtained in the first throw, the number obtained in the second throw, and the sum of these numbers, respectively, then X_1, X_2, and X are real random variables represented by

$$X_1(i, j) = i, \qquad X_2(i, j) = j, \qquad X(i, j) = i + j.$$

Let $X = X(\omega)$ be a real random variable on (Ω, P). The image $X(\Omega)$, the set of all the values $X(\omega)$ can take, is called the sample space of the random variable X, written Ω^X. For any set $B \subset \Omega^X$ the probability of the event that the value of X lies in B is equal to $P\{\omega | X(\omega) \in B\}$, that is, $P(X^{-1}(B))$. Viewing this as a function of a set $B \subset \Omega^X$, we denote it by $P^X(B)$. It is easy to check that P^X satisfies the conditions (p.1), (p.2), and (p.3) in Section 1.1 using the relations

$$X^{-1}(B_1 + B_2) = X^{-1}(B_1) + X^{-1}(B_2), \qquad X^{-1}(\Omega^X) = \Omega.$$

The probability measure P^X is called the *probability distribution* of X. In particular, we have

$$P^X\{b\} = P(X^{-1}(b)), \qquad b \in \Omega^X.$$

Since P^X is a probability measure, we have

$$P^X(B) = \sum_{b \in B} P^X\{b\},$$

which shows that P^X is completely determined by assigning $P^X\{b\}$ for every $b \in \Omega^X$. (Ω^X, P^X) is called the *probability space* of the random variable X.

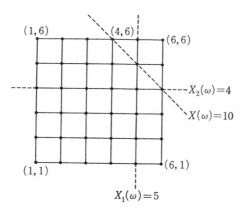

Figure 1.1

Let us determine the probability spaces of the random variables X_1, X_2, and X in the example of throwing a die twice, mentioned above.

$$\Omega^{X_1} = \Omega^{X_2} = \{1,2,3,4,5,6\}, \qquad \Omega^X = \{2,3,4,\ldots,12\},$$

$$P^{X_i}\{k\} = P\left(X_i^{-1}(k)\right) = \frac{\#X_i^{-1}(k)}{36} = \frac{6}{36} = \frac{1}{6}, \qquad i = 1,2,$$

$$P^X\{k\} = P\left(X^{-1}(k)\right) = \frac{\#X^{-1}(k)}{36} = \begin{cases} (k-1)/36 & (2 \le k \le 7), \\ (13-k)/36 & (8 \le k \le 12), \end{cases}$$

as we can see in Figure 1.1.

Let $X_1(\omega)$ and $X_2(\omega)$ be real random variables. Then the pair $X(\omega) = (X_1(\omega), X_2(\omega))$ is a function on Ω with values in $\mathbb{R}^2$, called a (two-dimensional) *vector-valued random variable* or simply a *random vector*. The sample space and the probability law of X are defined as follows:

$$\Omega^X = X(\Omega),$$

$$P^X(B) = P\{\omega | X(\omega) \in B\} = P\left(X^{-1}(B)\right), \qquad B \subset \Omega^X.$$

Ω^X is a subset of $\mathbb{R}^2$ and P^X is a probability measure on Ω^X. It is obvious that

$$\Omega^X \subset \Omega^{X_1} \times \Omega^{X_2}.$$

Note that these two sets do not always coincide with each other. $X_1(\omega)$ and $X_2(\omega)$ are *component variables* of $X(\omega)$, and $X(\omega)$ is called the *joint variable* of $X_1(\omega)$ and $X_2(\omega)$. The probability law P^X is often called the *joint probability distribution* of $X_1(\omega)$ and $X_2(\omega)$. For any natural number n we can define n-dimensional random vectors similarly.

The map that carries $(x_1, x_2, \ldots, x_n) \in \mathbb{R}^n$ to its kth component x_k is called the *k-projection*, written π_k. If $X(\omega)$ is the joint variable of $X_1(\omega), X_2(\omega), \ldots, X_n(\omega)$, then we have

$$X_k(\omega) = \pi_k(X(\omega)), \qquad \text{i.e., } X_k = \pi_k \circ X,$$

where the small circle $\circ$ denotes composition of maps.

Let $X(\omega)$ be a real random variable. For any real-valued function φ defined on Ω^X, we obtain a real random variable $Y(\omega)$ defined by

$$Y(\omega) = \varphi(X(\omega)) \qquad (\text{i.e., } Y = \varphi \circ X).$$

The probability space of Y is given by

$$\Omega^Y = (\varphi \circ X)(\Omega),$$

$$P^Y(C) = P\big((\varphi \circ X)^{-1}(C)\big).$$

A real random variable $Y(\omega)$ thus obtained from $X(\omega)$ is called a function of $X(\omega)$. Similarly an m-dimensional random vector $Y(\omega)$ is defined to be a function of an n-dimensional random vector $X(\omega)$ if there exists a map $\varphi \colon \Omega^X(\subset \mathbb{R}^n) \to \mathbb{R}^m$ such that $Y(\omega) = \varphi(X(\omega))$ for every ω. For example the kth component $X_k(\omega)$ of $X(\omega) = (X_1(\omega), X_2(\omega), \ldots, X_n(\omega))$ is a function of $X(\omega)$, since $X_k(\omega) = \pi_k(X(\omega))$.

Theorem 1.2.1. If $Y(\omega) = \varphi(X(\omega))$, then

$$\Omega^Y = \varphi(\Omega^X), \qquad P^Y(C) = P^X(\varphi^{-1}(C)).$$

Proof: Observing that $Y(\omega) = (\varphi \circ X)(\omega)$ and that

$$\omega \in (\varphi \circ X)^{-1}(C) \Leftrightarrow (\varphi \circ X)(\omega) \in C \Leftrightarrow \varphi(X(\omega)) \in C$$
$$\Leftrightarrow X(\omega) \in \varphi^{-1}(C) \Leftrightarrow \omega \in X^{-1}(\varphi^{-1}(C)),$$

we obtain

$$Y(\omega) = \varphi(X(\omega)) = (\varphi \circ X)(\omega)$$
$$\Omega^Y = (\varphi \circ X)(\Omega) = \varphi(X(\Omega)) = \varphi(\Omega^X),$$
$$P^Y(C) = P\big((\varphi \circ X)^{-1}(C)\big) = P\big(X^{-1}(\varphi^{-1}(C))\big) = P^X(\varphi^{-1}(C)). \qquad \blacksquare$$

For a real random variable X the *mean value* (or *expectation*) of X, written EX, is defined by

$$EX = \sum_{\omega \in \Omega} X(\omega) P\{\omega\}.$$

For any set $A \subset \Omega$ we denote

$$\sum_{\omega \in A} X(\omega) P\{\omega\}$$

by $E(X, A)$. The mean value of a random vector $X(\omega) = (X_1(\omega), X_2(\omega), \ldots, X_n(\omega))$ is defined by

$$EX = (EX_1, EX_2, \ldots, EX_n) \in \mathbb{R}^n.$$

Theorem 1.2.2. Let X and Y be random vectors. Then we have the following:

(i) (*additivity of mean values*)

$$E(aX + bY) = aEX + bEY \qquad (a, b: \text{constant})$$

(ii) $E(X, \sum_{i=1}^n A_i) = \sum_{i=1}^n E(X, A_i)$

(iii) If $X(\omega) = a$ (a: constant vector) on A, then $E(X, A) = P(A)a$. In particular, if $X(\omega) = a$ on Ω, then $EX = a$.

(iv) $EX = \sum_{x \in \Omega^X} P^X\{x\}x$

(v) If $Y(\omega) = \varphi(X(\omega))$ on Ω, then

$$EY = \sum_{x \in \Omega^X} \varphi(x) P^X\{x\}.$$

Let X and Y be real random variables, then we have the following:

(vi) $X(\omega) \geq Y(\omega) \Rightarrow EX \geq EY$, $E(X, A) \geq E(Y, A)$.

(vii) $X(\omega) \geq 0$, $A \subset B \Rightarrow E(X, A) \leq E(X, B)$.

Proof: We will only prove (iv) and (v) here, because the other statements are trivial.

(iv) Setting $A_x = X^{-1}(x)$, we have

$$\Omega = \sum_{x \in \Omega^X} A_x,$$

so

$$E(X) = E(X, \Omega) = \sum_{x \in \Omega^X} E(X, A_x)$$

by (ii). Since $X(\omega) = x$ on A_x, we have

$$E(X, A_x) = P(A_x)x = P^X\{x\}x.$$

Hence we obtain (iv) at once.

(v) In the same way as above, we have

$$E(\varphi(X)) = \sum_x E(\varphi(X), A_x) = \sum_x \varphi(x) P(A_x)$$

$$= \sum_x \varphi(x) P^X\{x\}$$

where x runs over Ω^X. ∎

Let A be a subset of Ω. The *indicator* of A, written 1_A, is defined by

$$1_A = \begin{cases} 1, & \omega \in A \\ 0, & \omega \in A^c. \end{cases}$$

This is a random variable that takes 1 or 0 according to whether A occurs or not. It is obvious that

$$1_\Omega(\omega) = 1 \qquad \text{for every } \omega \in \Omega.$$

$1_\Omega(\omega)$ is a random variable that takes 1 always, whereas 1 is just a fixed number. However, we often denote $1_\Omega(\omega)$ simply by 1 and accordingly $a \cdot 1_\Omega(\omega)$ (a: constant) simply by a unless there is any possibility of confusion. It is obvious that

$$E1_A = P(A), E(X, A) = E(X1_A).$$

Theorem 1.2.3. Let $e, e', e_1, e_2, \ldots, e_n$ denote the indicators of $A, A^c, A_1, A_2, \ldots, A_n$, respectively.

(i) $e' = 1 - e$.

(ii) If $A = \bigcap_{i=1}^n A_i$, then

$$e = \prod_{i=1}^n e_i \qquad \text{and} \qquad e = \min(e_1, e_2, \ldots, e_n).$$

(iii) If $A = \bigcup_{i=1}^n A_i$, then

$$e = 1 - \prod_{i=1}^n (1 - e_i) \qquad \text{and} \qquad e = \max(e_1, e_2, \ldots, e_n).$$

Proof: Assertions (i) and (ii) and the second part of (iii) are obvious. To prove the first part of (iii), observe the following:

$$A^c = \bigcap_{i=1}^n A_i^c \qquad \text{by de Morgan's law,}$$

$$1 - e = \prod_{i=1}^n (1 - e_i) \qquad \text{by (i) and (ii).} \qquad ∎$$

The *variance* of X, written $V(X)$, is defined as follows:

$$V(X) = E((X - EX)^2).$$

Setting $\varphi(x) = (x - EX)^2$ in Theorem 1.2.2(v), we have

$$V(X) = \sum_{x \in \Omega^X} (x - EX)^2 P^X\{x\}.$$

The *covariance* of X and Y, written $V(X, Y)$, is defined as

$$V(X, Y) = E((X - EX)(Y - EY)).$$

Replacing X by (X, Y), x by (x, y), and $\varphi(x)$ by $(x - EX)(y - EY)$ in Theorem 1.2.2(v), we obtain

$$V(X, Y) = \sum_{(x, y) \in \Omega^{(X, Y)}} (x - EX)(y - EY) P^{(X, Y)}\{(x, y)\}.$$

The following properties are verified easily:

$$V(X, Y) = V(Y, X),$$

$$V(X, a) = 0 \qquad (a: \text{constant}),$$

$$V(X, X) = V(X) \geq 0.$$

Theorem 1.2.4.
    (i) $V(aX + b) = a^2 V(X)$, $\qquad V(aX + b, cY + d) = ac V(X, Y)$.
    (ii) $V(aX + bY) = a^2 V(X) + 2ab V(X, Y) + b^2 V(Y)$.
    (iii) $V(X) = EX^2 - (EX)^2$, $\qquad V(X, Y) = E(XY) - EXEY$.
    (iv) $|V(X, Y)| \leq \sqrt{V(X)V(Y)}$.

Proof: To prove (i), (ii), and (iii), take the mean values of both sides of the following:

$$((aX + b) - E(aX + b))^2 = a^2(X - E(X))^2,$$

$$((aX + b) - E(aX + b))((cY + d) - E(cY + d))$$
$$= ac(X - EX)(Y - EY),$$

$$((aX + bY) - E(aX + bY))^2 = a^2(X - EX)^2 + 2ab(X - EX)$$
$$\times (Y - EY) + b^2(Y - EY)^2,$$

$$(X - EX)^2 = X^2 - 2(EX)X + (EX)^2,$$

$$(X - EX)(Y - EY) = XY - (EX)Y - (EY)X + EXEY.$$

To prove (iv), observe that

$$0 \leq V(tX + Y) = E((tX + Y) - E(tX + Y))^2$$
$$= E\left(t^2(X - EX)^2 + 2t(X - EX)(Y - EY) + (Y - EY)^2\right)$$
$$= t^2V(X) + 2tV(X, Y) + V(Y).$$

The last quadratic function of t is nonnegative for every real value of t. Hence

$$V(X, Y)^2 - V(X)V(Y) \leq 0, \quad \text{i.e., } |V(X, Y)| \leq \sqrt{V(X)V(Y)}.$$

∎

The *standard deviation* of X, $\sigma(X)$ and the *correlation coefficient* of X and Y, $R(X, Y)$ are defined as follows:

$$\sigma(X) = \sqrt{V(X)},$$

$$R(X, Y) = \frac{V(X, Y)}{\sigma(X)\sigma(Y)} \qquad \text{if } \sigma(X)\sigma(Y) > 0.$$

A condition on real random variables $X(\omega)$, $Y(\omega)$, and $Z(\omega)$, for example, $X(\omega) + Y(\omega) \geq Z(\omega)$, is also regarded as a condition on ω. Hence it is an event, and the probability of occurrence of this event is

$$P\{\omega | X(\omega) + Y(\omega) \geq Z(\omega)\},$$

which is denoted simply by

$$P\{X(\omega) + Y(\omega) \geq Z(\omega)\} \qquad \text{or} \qquad P\{X + Y \geq Z\}.$$

If the probability of occurrence of α equals 1; that is, $P(\alpha) = 1$, we say that $\alpha(\omega)$ holds (or occurs) *almost surely* or that $\alpha(\omega)$ holds with probability 1, written $\alpha(\omega)$ a.s. Mean values, variances, covariances, and correlation coefficients are the same for the random variables equal to each other almost surely:

$$X(\omega) = X'(\omega) \text{ a.s.} \Rightarrow EX = EX', \, V(X) = V(X'), \, \sigma(X) = \sigma(X'),$$
$$X(\omega) = X'(\omega) \text{ a.s.}, \, Y(\omega) = Y'(\omega) \text{ a.s.} \Rightarrow V(X, Y) = V(X', Y'),$$
$$R(X, Y) = R(X', Y').$$

To prove these properties it suffices to show that

$$X(\omega) = X'(\omega) \text{ a.s.} \Rightarrow EX = EX'.$$

Let A denote the set $\{\omega | X(\omega) \neq X'(\omega)\}$. Then $P(A) = 0$, so $P\{\omega\} = 0$ for every $\omega \in A$. This implies that $E(X, A) = E(X', A) = 0$. Since $X(\omega) = X'(\omega)$ on A^c, we have $E(X, A^c) = E(X', A^c)$. Hence we have

$$E(X) = E(X, A) + E(X, A^c) = E(X', A) + E(X', A^c) = E(X').$$

Theorem 1.2.5. If $X(\omega) \geq 0$ and $EX = 0$, then $X(\omega) = 0$ a.s.

Proof: It suffices to show that $P(A) = 0$ for $A = \{\omega \mid X(\omega) > 0\}$. Suppose that $P(A) > 0$. Then we can find a point $\omega_0 \in A$ such that $P\{\omega_0\} > 0$. Since $\omega_0 \in A$, $X(\omega_0)$ is positive. Hence we have

$$EX = \sum_{\omega \in \Omega} X(\omega) P\{\omega\} \geq X(\omega_0) P\{\omega_0\} > 0,$$

contrary to the assumption.

Theorem 1.2.6.
    (i) $\sigma(X) = 0 \Leftrightarrow X(\omega) = EX$ a.s.
    (ii) $-1 \leq R(X, Y) \leq 1$
    (iii) If $R(X, Y) = \pm 1$, then Y is a linear function of X given by the following formula:

$$Y(\omega) - EY = \pm \frac{\sigma(Y)}{\sigma(X)} (X(\omega) - EX) \text{ a.s.}$$

The signs $\pm$ in the assumption correspond to those in the conclusion, respectively.

Proof:
    (i) Apply Theorem 1.2.5 to the variable $(X - EX)^2$.
    (ii) Obvious by Theorem 1.2.4(iv).
    (iii) Set $k = \sigma(Y)/\sigma(X)$ and use Theorem 1.2.4(ii) to check

$$\sigma(Y - kX)^2 = \sigma(Y)^2 - 2k\sigma(X)\sigma(Y)R(X, Y) + k^2\sigma(X)^2$$

$$= \sigma(Y)^2 - 2\sigma(Y)^2 R(X, Y) + \sigma(Y)^2$$

$$= 2\sigma(Y)^2(1 - R(X, Y)),$$

which implies that

$$R(X, Y) = 1 \Leftrightarrow \sigma(Y - kX) = 0$$

$$\Leftrightarrow Y(\omega) - kX(\omega) = EY - kEX \text{ a.s.}$$

$$\Leftrightarrow Y(\omega) - EY = k(X(\omega) - EX) \text{ a.s.},$$

which proves the case of sign $+$. Similarly we can prove the case of sign $-$ by setting $k = -\sigma(Y)/\sigma(X)$. ∎

Theorem 1.2.7 (Chebyshev's inequality). For every $a > 0$ we have the following inequality:

$$P\{|X(\omega) - EX| > a\sigma(X)\} \leq 1/a^2;$$

that is,

$$P\{EX - a\sigma(X) \leq X(\omega) \leq EX + a\sigma(X)\} \leq 1 - 1/a^2.$$

Proof: If $\sigma(X) = 0$, then $X(\omega) = EX$ a.s. by (i) of the last theorem, so our inequality holds obviously. Suppose that $\sigma(X) > 0$. Denoting by A the event in consideration and noting that $(X - EX)^2 \geq 0$ on Ω, we obtain

$$\sigma(X)^2 = E(X - EX)^2 \geq E((X - EX)^2, A) \geq a^2\sigma(X)^2 P(A),$$

which implies that $P(A) \leq 1/a^2$. ∎

Exercise 1.2. X and Y are real random variables except in (ii).

(i) Prove that $|EX| \leq E|X|$ and $|EX| \leq \sqrt{EX^2}$
   [*Hint*: Derive the first inequality from $-|X| \leq X \leq |X|$ and the second one from $EX^2 - (EX)^2 = V(X)$.]

(ii) Use the second inequality of (i) to prove

$$\|EX\| \leq E\|X\|,$$

where X is a random vector and $\| \ \|$ denotes the Euclidean norm.

(iii) Use $E1_A = P(A)$ and Theorem 1.2.3 to prove inclusion–exclusion formula (i) of Exercise 1.1 and use $P(A^c) = 1 - P(A)$ to derive from this the second formula (ii), noting that

$$\sum_{k=0}^{n} (-1)^k \binom{n}{k} = (1 - 1)^n = 0.$$

(iv) Prove that $R(X, aX + b) = \text{sgn } a$ where $\sigma(X) > 0$, $a \neq 0$ and sgn $a = 1$ ($a > 0$) and -1 ($a < 0$).

(v) Use Chebyshev's inequality to prove that

$$P\left\{\left|\frac{Y(\omega) - EY}{\sigma(Y)} - \frac{X(\omega) - EX}{\sigma(X)}\right| > a\sqrt{2(1 - R(X, Y))}\right\}$$

$$\leq \frac{1}{a^2} \quad (a > 0),$$

where we assume that $\sigma(X), \sigma(Y) > 0$.

(vi) Suppose that we throw a die twice and denote by X_1 (X_2) the number obtained in the first (second) throw. Express

$$R(a_1 X_1 + a_2 X_2, b_1 Y_1 + b_2 Y_2)$$

in terms of a_1, a_2, b_1, and b_2.

(vii) Express the minimum value of $f(t) = E(X - t)^2$ ($-\infty < t < \infty$) and the value of t minimizing $f(t)$ in terms of EX and $V(X)$.

(viii) Express the minimum value of $g(t, s) = E(Y - tX - s)^2$ ($-\infty < t, s < \infty$) and the values of t and s minimizing $g(t, s)$ in terms of EX, EY, $V(X)$, $V(Y)$, and $V(X, Y)$.

1.3   Mixing, direct composition, and tree composition

1.3a   General random variables

In the last section we defined real random variables and random vectors. Similarly we can consider *general random variables*, which may take on values other than real numbers or vectors. Suppose that we observe the number obtained when throwing a die. The probability space for this trial is given by

$$\Omega = \{1, 2, 3, 4, 5, 6\}, \qquad P(A) = \#A/6,$$

as we have already seen. Now define $X(\omega)$ by

$$X(\omega) = \begin{cases} \text{odd} & (\omega = 1, 3, 5) \\ \text{even} & (\omega = 2, 4, 6). \end{cases}$$

This is regarded as a random variable with values in {odd, even} and indicates whether the number obtained is odd or even. The sample space Ω^X and the probability law P^X for this random variable are given as follows:

$$\Omega^X = \{\text{odd}, \text{even}\},$$

$$P^X\{\text{odd}\} = P\{\omega | X(\omega) \text{ is odd}\} = P\{1, 3, 5\} = \tfrac{1}{2}$$

$$P^X\{\text{even}\} = P\{\omega | X(\omega) \text{ is even}\} = P\{2, 4, 6\} = \tfrac{1}{2}.$$

Suppose that we observe the number on a card drawn from a box with 10 cards numbered $1, 2, \ldots, 10$. The probability space for this trial, (Ω, P) is given by

$$\Omega = \{1, 2, \ldots, 10\}, \qquad P(A) = \#A/10.$$

We also assume that cards 1 and 2 are red, cards 3, 4, and 5 are blue, and the other five cards are white. Let Y denote the color of the card drawn. Then Y is represented as follows:

$$Y(1) = Y(2) = \text{red},$$

$$Y(3) = Y(4) = Y(5) = \text{blue},$$

$$Y(6) = Y(7) = \cdots = Y(10) = \text{white}.$$

This is regarded as a random variable whose sample space and probability law are given as follows:

$$\Omega^Y = \{\text{red, blue, white}\},$$

$$P^Y\{\text{red}\} = 2/10 = 1/5, \qquad P^Y\{\text{blue}\} = 3/10,$$

$$P^Y\{\text{white}\} = 5/10 = 1/2.$$

As in the case of real random variables or random vectors, the probability space of a general random variable is defined as the pair of the sample space and the probability law, but the mean value of a general random variable cannot always be defined.

1.3b   Mixing of a trial

Let T be a trial, (Ω, P) the sample space for T, and $X(\omega)$ a random variable on (Ω, P). If we observe every outcome $\omega \in \Omega$, paying attention only to the value of $X(\omega)$, we obtain a new trial, which is called the *mixing* of T by $X(\omega)$, written T_X. For example, we encounter a mixing when we pay attention to "even or odd" of the number obtained when throwing a die or to the color of the cards drawn from a box. The sample space of T_X obviously coincides with Ω^X, the sample space of X. A subset B of Ω^X is regarded as an event in the trial T_X, which in turn is represented by the set $X^{-1}(B)(\subset \Omega)$ as an event on the original trial T. The probability of this event is $P(X^{-1}(B))$, that is, $P^X(B)$. Hence the probability measure for T_X coincides with P^X, the probability distribution of X. Thus the probability space for T_X coincides with that of X, (Ω^X, P^X). The term "mixing" comes from the fact that any outcomes of T are identified in T_X if and only if they correspond to the same value of X.

Let φ be a map from Ω^X into a general set Φ. Then $\varphi(x)$ is regarded as a random variable on (Ω^X, P^X). Let $Y(\omega)$ denote $\varphi(X(\omega))$. $Y(\omega)$ is regarded as a random variable on (Ω, P). Since $Y(\omega)$ takes $\varphi(x)$ when $X(\omega)$ takes x, $Y(\omega)$ represents on (Ω, P) what $\varphi(x)$ does on (Ω^X, P^X). Therefore it is reasonable that the probability distribution of $\varphi(x)$ on (Ω^X, P^X), written $(P^X)^\varphi$, coincides with that of $Y(\omega)$ on (Ω, P), that is,

P^Y. Similarly for the mean value of $\varphi(x)$ on (Ω^X, P^X), $E^X\varphi$, in case $\Phi \subset \mathbb{R}^n$. In fact, these are guaranteed by the following:

Theorem 1.3.1.
    (i) $P^Y(\Lambda) = P^X(\varphi \in \Lambda)$ $(\Lambda \subset \Omega^Y = \varphi(\Omega^X))$.
    (ii) $EY = E^X\varphi$ in case $\Phi \subset \mathbb{R}^n$.

Proof:
    (i) $P\{Y \in \Lambda\} = P(Y^{-1}(\Lambda)) = P((\varphi \circ X)^{-1}(\Lambda)) = P(X^{-1}(\varphi^{-1}(\Lambda)))$
$$= P^X(\varphi^{-1}(\Lambda)) = P^X\{\varphi \in \Lambda\}.$$
    (ii) Use Theorem 1.2.2(v) and the definition of E^X. ∎

Since variances, covariances, standard deviations, and correlation coefficients are defined in terms of mean values, the second proposition of the theorem above implies that if
$$Y(\omega) = \varphi(X(\omega)), \qquad Z(\omega) = \psi(X(\omega)),$$
then
$$V(Y) = V^X(\varphi), \qquad V(Y, Z) = V^X(\varphi, \psi),$$
$$\sigma(Y) = \sigma^X(\varphi), \qquad R(Y, Z) = R^X(\varphi, \psi).$$

1.3.c   *Direct composition of trials*

Let T_1 and T_2 be two trials whose probability spaces are (Ω_1, P_1) and (Ω_2, P_2), respectively, and denote the composite trial of T_1 and T_2 by $\tilde{T}$. We call the trial $\tilde{T}$ the direct composition of T_1 and T_2, written $T_1 \times T_2$, to distinguish it from the tree composition explained later. Since every outcome of $\tilde{T}$ is represented by a pair of one outcome of T_1 and one of T_2, the sample space of $\tilde{T}$, written $\tilde{\Omega}$, is given by
$$\tilde{\Omega} = \{(\omega_1, \omega_2) | \omega_1 \in \Omega_1, \omega_2 \in \Omega_2\}$$
$$= \Omega_1 \times \Omega_2 \qquad \text{(Cartesian product)}.$$

We will discuss the probability measure $\tilde{P}$ of the composite trial $\tilde{T}$. Let $\pi_i: \Omega_1 \times \Omega_2 \to \Omega_i$ denote the ith projection carrying $\tilde{\omega} = (\omega_1, \omega_2)$ to ω_i. If $\tilde{\omega} = (\omega_1, \omega_2)$ is an income of $\tilde{T}$, then $\omega_i = \pi_i(\tilde{\omega})$ is the outcome of T_i corresponding to $\tilde{\omega}$. In other words T_i is the mixing of $\tilde{T}$ by $\pi_i(\tilde{\omega})$. Hence the following relations must hold:
$$\tilde{P}\pi_i^{-1} = P_i, \qquad i = 1, 2;$$
that is,
$$\tilde{P}(\pi_i^{-1}(A_i)) = P_i(A_i) \qquad (A_i \subset \Omega_i; i = 1, 2).$$
There are many probability measures $\tilde{P}$ satisfying these relations. For

example, toss a coin twice and observe the faces (heads or tails). Such a trial $\tilde{T}$ is the composition of the trials T_1 and T_2 corresponding to the first toss and the second toss, respectively. Denoting heads by 1 and tails by 0, we obtain the following:

$$\Omega_i = \{0,1\}$$

$$P_i\{j\} = \tfrac{1}{2}, \qquad j = 0,1,$$

$$\tilde{\Omega} = \{(0,0),(0,1),(1,0),(1,1)\},$$

and the probability measure $\tilde{P}$ on $\tilde{\Omega}$ satisfying the conditions above is given as follows:

$$\tilde{P}\{(0,0)\} = \alpha, \qquad \tilde{P}\{(0,1)\} = \tfrac{1}{2} - \alpha,$$

$$\tilde{P}\{(1,0)\} = \tfrac{1}{2} - \alpha, \qquad \tilde{P}\{(1,1)\} = \alpha,$$

where α is a parameter ranging over $[0, \tfrac{1}{2}]$.

Now we will return to the general case. In order to determine the probability law of $\tilde{T}$ we have to select one from among all probability measures $\tilde{P}$ satisfying $\tilde{P}\pi_i^{-1} = P_i$ $(i = 1, 2)$. As a principle of such a selection we use the *multiplicative law*, $\tilde{P}\{(\omega_1, \omega_2)\} = P_1\{\omega_1\}P_2\{\omega_2\}$, which implies

$$\tilde{P}(\tilde{A}) = \sum_{(\omega_1, \omega_2) \in \tilde{A}} P_1\{\omega_1\}P_2\{\omega_2\}$$

by Theorem 1(v). To see that $\tilde{P}$ satisfies the conditions above, it suffices to observe the following:

$$\tilde{P}(\tilde{\Omega}) = \tilde{P}(\Omega_1 \times \Omega_2) = P_1(\Omega_1)P_2(\Omega_2) = 1,$$

$$\tilde{P}(\pi_1^{-1}(A_1)) = \tilde{P}(A_1 \times \Omega_2) = P_1(A_1)P_2(\Omega_2) = P_1(A_1),$$

$$\tilde{P}(\pi_2^{-1}(A_2)) = P_2(A_2).$$

In the example mentioned above, this probability measure $\tilde{P}$ corresponds to $\alpha = \tfrac{1}{4}$.

For any probability measures P_1 and P_2 defined on Ω_1 and Ω_2, respectively, the probability measure $\tilde{P}$ on $\Omega_1 \times \Omega_2$ obtained from P_1 and P_2 by the multiplicative law is called the *direct product* of P_1 and P_2, written $P_1 \times P_2$, and the probability space $(\Omega_1 \times \Omega_2, P_1 \times P_2)$ is called the *direct product* of (Ω_1, P_1) and (Ω_2, P_2), written $(\Omega_1, P_1) \times (\Omega_2, P_2)$. Using this terminology we say that the probability space for the direct composition of two trials T_1 and T_2 is the direct product of the probability spaces for T_1 and T_2.

In the discussion above we have used the multiplicative law to determine the probability measure of the direct composition. We should admit this law as an axiom, or, equivalently, we may define the direct composition of T_1 and T_2 to be a trial whose probability space is the direct product of the probability spaces of T_1 and T_2.

Instead of the multiplicative law, we can use the *principle of maximum entropy*. Let P be a probability measure on $\Omega = \{a_1, a_2, \ldots, a_m\}$. The *entropy* of P, $\varepsilon(P)$ is defined as follows:

$$\varepsilon(P) = \sum_{i=1}^{m} P\{a_i\} \log \frac{1}{P\{a_i\}},$$

where we define

$$0 \log \frac{1}{0} = \lim_{x \to 0} x \log \frac{1}{x} = 0$$

by convention. Let $\mathbf{P}$ be a given family of probability measures on Ω. The principle of maximum entropy is to choose from among $\mathbf{P}$ the probability measure that maximizes $\varepsilon(P)$. We will apply this principle to the direct composition discussed above. Let $\tilde{P}$ be the class of all $\tilde{\mathbf{P}}$ on $\Omega_1 \times \Omega_2$ satisfying

$$\tilde{P}\pi_1^{-1} = P_1, \qquad \tilde{P}\pi_2^{-1} = P_2,$$

where P_1 and P_2 are given probability measures on Ω_1 and Ω_2, respectively. We want to find the $\tilde{P}$ that maximizes $\varepsilon(\tilde{P})$. Set

$$\Omega_1 = \{a_1, a_2, \ldots, a_n\}, \qquad \Omega_2 = \{b_1, b_2, \ldots, b_m\}$$

and

$$p_i = P_1\{a_i\}, \qquad q_j = P_2\{b_j\}, \qquad r_{ij} = \tilde{P}\{(a_i, b_i)\}.$$

Then the conditions above are expressed as follows:

$$\text{(c)} \quad \begin{aligned} \sum_{j=1}^{m} r_{ij} &= p_i, \qquad i = 1, 2, \ldots, n, \\ \sum_{i=1}^{n} r_{ij} &= q_j, \qquad j = 1, 2, \ldots, m, \end{aligned}$$

and the entropy $\varepsilon(\tilde{P})$ is given by

$$\varepsilon(\tilde{P}) = \sum_{i,j} r_{ij} \log \frac{1}{r_{ij}}.$$

To maximize $\varepsilon(\tilde{P})$ under conditions (c) we will use Lagrange's method of indeterminate coefficients as follows:

$$F(r_{11}, r_{12}, \ldots, r_{lm}) = \sum_{i,j} r_{ij} \log \frac{1}{r_{ij}} - \sum_i \alpha_i \left(\sum_j r_{ij} - p_i \right) - \sum_j \beta_j \left(\sum_i r_{ij} - q_j \right)$$

$$\frac{\partial F}{\partial r_{ij}} = \log \frac{1}{r_{ij}} - 1 - \alpha_i - \beta_j = 0,$$

$$r_{ij} = e^{-1-\alpha_i-\beta_j}.$$

This, combined with conditions (c), implies

$$p_i = e^{-1} e^{-\alpha_i} \sum_k e^{-\beta_k}, \qquad q_j = e^{-1} e^{-\beta_j} \sum_l e^{-\alpha_l},$$

so

$$\frac{r_{ij}}{p_i q_j} = \frac{1}{e^{-1} \sum_k e^{-\alpha_k} \sum_l e^{-\beta_l}}$$

Denoting the right-hand side by γ we obtain

$$1 = \sum_{i,j} r_{ij} = \gamma \sum_i p_i \sum_j q_j = \gamma.$$

This proves that $r_{ij} = p_i q_j$, the same result we obtained by using the multiplicative law.

A similar observation can be made for the direct composition of a finite number of trials, and the probability space for the composite trial is the direct product of those for the component trials. Obviously we have the following:

Theorem 1.3.2. If $\tilde{P} = P_1 \times P_2 \cdots \times P_n$, that is,

$$\tilde{P}\{(\omega_1, \omega_2, \ldots, \omega_n)\} = P_1\{\omega_1\} P_2\{\omega_2\} \ldots P_n\{\omega_n\},$$

then

$$\tilde{P}(A_1 \times A_2 \times \cdots \times A_n) = P_1(A_1) P_2(A_2) \ldots P_n(A_n).$$

1.3.d   Tree composition of trials

By superposing direct composition and mixing we can make a new trial. To illustrate this we will mention two examples.

Example 1.3.1. Suppose that two players A and B toss a coin interchangeably with A starting. The one who gets heads first will be the winner of this game. If neither A nor B gets heads up to the 10th toss (i.e., up to the 5th toss for each player), the game will end in a draw. We regard the whole process of this game as a trial and denote it by T. The sample space of T, Ω, consists of the following 11 sample points where 001, for example, means that A gets tails, B gets tails, and A gets heads–the game ending up with A winning.

$$a_1 = 1 \qquad\qquad (A \text{ wins})$$
$$a_2 = 01 \qquad\qquad (B \text{ wins})$$
$$a_3 = 001 \qquad\qquad (A \text{ wins})$$
$$\vdots$$
$$a_{10} = 0000000001 \qquad (B \text{ wins})$$
$$a_{11} = 0000000000 \qquad (\text{draw}).$$

It is known that the probability law of T, P, is given by

$$P\{a_k\} = 2^{-k} \ (k = 1, 2, \ldots, 10), \qquad P\{a_{11}\} = 2^{-10}.$$

We will explain the reason for these equalities' holding by superposition of direct composition and mixing. If the sample point a_k $(k \le 10)$ is realized, the game ends up with A or B winning at the kth toss according to whether k is odd or even. But we imagine that the players continue to take turns up to the 10th toss even if the game is over. Then we get a large trial, which will be denoted by $\tilde{T}$. Since $\tilde{T}$ is the direct composition of 10 trials, in each of which A or B toss a coin, the probability space for $\tilde{T}$, $(\tilde{\Omega}, \tilde{P})$, is given by

$$\tilde{\Omega} = \{0, 1\}^{10} = \{(i_1, i_2, \ldots, i_{10}) \,|\, i_1, i_2, \ldots, i_{10} = 0, 1\},$$

$$\tilde{P}\{(i_1, i_2, \ldots, i_{10})\} = 2^{-10}$$

Define a map $X \colon \tilde{\Omega} \to \Omega$ by

$$X(i_1, i_2, \ldots, i_{10}) = \begin{cases} a_1 & (i_1 = 1) \\ a_2 & (i_1 = 0, i_2 = 1) \\ \vdots & \\ a_{10} & (i_1 = i_2 = \cdots = i_9 = 0, i_{10} = 1) \\ a_{11} & (i_1 = i_2 = \cdots = i_{10} = 0). \end{cases}$$

$X(\tilde{\omega})$ is an Ω-valued random variable. It is obvious that the trial T mentioned above is obtained by mixing of $\tilde{T}$ by $X(\tilde{\omega})$. Hence the probability law P coincides with the probability law of X, P^X so that

$$P\{a_k\} = \tilde{P}^X\{a_k\} = \tilde{P}\{X(\tilde{\omega}) = a_k\}$$

$$= \tilde{P}\{(i_1, i_2, \ldots, i_{10}) | i_1 = i_2 = \cdots = i_{k-1} = 0, i_k = 1\}$$

$$= 2^{10-k} \cdot 1/2^{10} = 2^{-k}$$

for $k = 1, 2, \ldots, 10$, and $P\{a_{11}\} = 2^{-10}$. The probability of A's winning is equal to

$$\sum_{k=1}^{5} P\{a_{2k-1}\} = \sum_{k=1}^{5} 2^{-(2k-1)} = \tfrac{2}{3}(1 - 2^{-10});$$

that of B's winning is equal to

$$\sum_{k=1}^{5} P\{a_{2k}\} = \sum_{k=1}^{5} 2^{-2k} = \tfrac{1}{3}(1 - 2^{-10}),$$

and the probability of draw is obviously 2^{-10}.

If we pay attention only to the winner, we have a simpler trial T', which is obtained by the mixing of T. The probability space for T', (Ω', P'), is given by

$$\Omega' = \{A \text{ wins}, B \text{ wins}, \text{draw}\}$$

$$P'\{A \text{ wins}\} = \tfrac{2}{3}(1 - 2^{-10}),$$

$$P'\{B \text{ wins}\} = \tfrac{1}{3}(1 - 2^{-10}),$$

$$P'\{\text{draw}\} = 2^{-10}.$$

Example 1.3.2. Suppose that we have a box B_0 that contains n cards numbered $1, 2, \ldots, n$. We draw a card from B_0, and then draw another card without replacement, and observe the pair of the numbers on these two cards. Let T denote this trial. We will show that the probability space for T, (Ω, P), is given by

$$\Omega = \{(i, j) | i, j = 1, 2, \ldots, n \ (i \neq j)\},$$

$$P\{(i, j)\} = 1/n(n - 1).$$

Let T_0 denote the first draw. If card i is drawn in T_0, then the second draw is from the box containing the cards numbered $1, 2, \ldots, i - 1$, $i + 1, \ldots, n$. This box is denoted by B_i, and a draw from B_i is denoted by T_i. The trial T to be observed is constructed in terms of $T_0, T_1, \ldots, T_n$ by the following scheme:

$$T_0 \begin{cases} \text{card 1 comes out} \to T_1 \\ \text{card 2 comes out} \to T_2 \\ \vdots \\ \text{card } n \text{ comes out} \to T_n \end{cases} \quad \text{or simply } T_0 \begin{matrix} T_1 \\ T_2 \\ \vdots \\ T_n \end{matrix}$$

In this sense, such a trial T is called a *tree composition*.

Let $\tilde{T}$ denote the direct composition of $T_0, T_1, T_2, \ldots, T_n$, that is, a trial of drawing a card from each box B_i, $i = 0, 1, 2, \ldots, n$, and observing the $(n + 1)$-tuple of numbers of the cards. Since the probability spaces for $T_0, T_1, T_2, \ldots, T_n$ are given, respectively, by

$$\Omega_0 = \{1, 2, \ldots, n\}, \qquad P_0\{j\} = 1/n$$
$$\Omega_i = \{1, 2, \ldots, i-1, i+1, \ldots, n\}, \qquad P_i\{j\} = 1/(n-1),$$

the probability space for $\tilde{T}$ is given by

$$\tilde{\Omega} = \Omega_0 \times \Omega_1 \times \cdots \times \Omega_n, \qquad \tilde{P} = P_0 \times P_1 \times \cdots \times P_n.$$

We can interpret T as follows. Having first drawn a card, say, i, we draw a card from each box B_α $(\alpha = 1, 2, \ldots, n)$ in any way and pay attention only to the card, say j, drawn from the box indexed with i, ignoring the other cards. Then (i, j) is regarded as an outcome of T. The outcome (i, j) of T corresponding to an outcome $\tilde{\omega}$ of $\tilde{T}$ is given by

$$i = \pi_0(\tilde{\omega}), \qquad j = \pi_i(\tilde{\omega}) = \pi_{\pi_0(\omega)}(\omega),$$

where π_k is the kth projection from $\tilde{\Omega}$ to Ω_k. If we define

$$X(\tilde{\omega}) = \left(\pi_0(\tilde{\omega}), \pi_{\pi_0(\tilde{\omega})}(\tilde{\omega})\right),$$

then T is regarded as the trial $\tilde{T}_X$ that is obtained from $\tilde{T}$ by mixing by $X(\tilde{\omega})$. Hence the probability space (Ω, P) for T is given as follows:

$$\Omega = \{(i, j) | i, j = 1, 2, \ldots, n \ (i \neq j)\},$$
$$\begin{aligned} P\{(i, j)\} &= \tilde{P}^X\{(i, j)\} = \tilde{P}\{X(\tilde{\omega}) = (i, j)\} \\ &= \tilde{P}\{\pi_0(\tilde{\omega}) = i, \pi_{\pi_0(\tilde{\omega})}(\tilde{\omega}) = j\} \\ &= \tilde{P}\{\pi_0(\tilde{\omega}) = i, \pi_i(\tilde{\omega}) = j\} \\ &= \tilde{P}\{\pi_0(\tilde{\omega}) = i, \pi_i(\tilde{\omega}) = j, \pi_k(\tilde{\omega}) \in \Omega_k \ (k \neq i, j)\} \\ &= P_0\{i\} P_i\{j\} \prod_{k \neq i, j} P_k(\Omega_k) \\ &= P_0\{i\} P_j\{j\} \\ &= \frac{1}{n} \cdot \frac{1}{n-1} = \frac{1}{n(n-1)}. \end{aligned}$$

For general trials we can define tree compositions similarly. Let $(\Omega_0 \equiv \{a_1, a_2, \ldots, a_n\}, P_0)$ denote the probability space for T_0. For each $a \in \Omega_0$ we are given a trial T_a with probability space (Ω_a, P_a). Define a tree composition T as follows:

$$T_0 \begin{cases} a_1 \text{ is realized} \rightarrow T_{a_1} \\ a_2 \text{ is realized} \rightarrow T_{a_2} \\ \vdots \\ a_n \text{ is realized} \rightarrow T_{a_n} \end{cases} \qquad \text{or simply } T_0 \begin{array}{l} \nearrow T_{a_1} \\ \nearrow T_{a_2} \\ \vdots \\ \searrow T_{a_n} \end{array}$$

As we saw in the example above, this trial T is obtained by superposition of direct composition and mixing. The sample space of T is

$$\Omega = \{(a, b) \mid a \in \Omega_0, b \in \Omega_a\},$$

and the probability measure of T is determined by the following:

Theorem 1.3.3 (Multiplicative law of tree composition).

$$P\{(a, b)\} = P_0\{a\} P_a\{b\}, \qquad (a, b) \in \Omega.$$

Proof: The probability space of $\tilde{T} = T_0 \times T_{a_1} \times \cdots \times T_{a_n}, (\tilde{\Omega}, \tilde{P})$ is given by

$$\tilde{\Omega} = \Omega_0 \times \Omega_{a_1} \times \Omega_{a_2} \times \cdots \times \Omega_{a_n}, \qquad \tilde{P} = P_0 \times P_{a_1} \times P_{a_2} \times \cdots \times P_{a_n}.$$

Since T is obtained by mixing of $\tilde{T}$ by the random variable

$$X(\tilde{\omega}) = \left(\pi_0(\tilde{\omega}), \pi_{\pi_0(\tilde{\omega})}(\tilde{\omega}) \right),$$

our theorem can be proved in the same way as the example above. ∎

It should be noted that the multiplicative law for direct composition was admitted as an axiom, whereas that for tree composition was proved as a theorem.

In case $T_1 = T_2 = \cdots = T_n$, the tree composition:

coincides with $T_0 \times T_1$. Hence the direct composition $T_0 \times T_1$ is a special

tree composition.

The tree composition discussed above consists of two steps. Similarly, we can consider tree compositions consisting of a finite number of steps as shown below:

We can regard the tree composition consisting of three steps as one of two steps by viewing the first two steps as a trial. Hence the multiplicative law for this case is as follows:

$$P\{(a, b, c)\} = P_0\{a\} P_a\{b\} P_{ab}\{c\}.$$

Exercise 1.3.

   (i) Let X and Y be the numbers of cards drawn in Example 1.3.2. Find the following: P^X, P^Y, EX, EY, $V(X)$, $V(Y)$, $\sigma(X)$, $\sigma(Y)$, $P^{(X,Y)}$, $V(X, Y)$, $R(X, Y)$.

  (ii) We have a box containing 18 cards, 2 of which are numbered i for $i = 1, 2, \ldots, 9$. We draw 4 cards one by one from this box, arrange them in the order of coming out, and make a number of 4 digits. What is the probability that this number exceeds 5283? [*Hint*: Denote by $P_0\{i\}$ the probability that card i is drawn from the box and by $P_i\{j\}$ the probability that card j is drawn from the remaining cards and define $P_{ij}\{k\}$ and $P_{ijk}\{l\}$ similarly. Then the probability in question is

$$P_0\{6, 7, 8, 9\} + P_0\{5\} P_5\{3, 4, \ldots, 9\} + P_0\{5\} P_5\{2\} P_{52}\{9\}$$

$$+ P_0\{5\} P_5\{2\} P_{52}\{8\} P_{528}\{4, 5, \ldots, 9\},$$

where

$$P_i(A) = \sum_{j \in A} P_i\{j\}$$

and similarly for P_{ij} and P_{ijk}.]

 (iii) Prove that the uniform probability measure $P\{a_i\} = 1/n$ on $\Omega = \{a_1, a_2, \ldots, a_n\}$ has the maximum entropy among all probability measures on Ω.

1.4   Conditional probabilities

Let (Ω, P) be a probability space and A and B any subsets of Ω, where $A \neq \varnothing$. If $P(A) > 0$, the (*conditional*) *probability* of B under A is defined to be equal to $P(A \cap B)/P(A)$ and is denoted by $P(B|A)$ or by $P_A(B)$. If $P(A) = 0$, we define the conditional probability by

$$P_A(B) = \begin{cases} 1, & \omega_0 \in B, \\ 0, & \omega_0 \notin B \end{cases}$$

for convention, where ω_0 is any fixed point in A.

Theorem 1.4.1.

(i) $P(A \cap B) = P(A) P_A(B)$.

(ii) For A fixed, $P_A(B)$ is a probability measure on Ω as a function of B such that $P_A(A) = 1$. (Hence P_A is called the (conditional) probability measure under A.)

Proof: Obvious by the definition. ∎

Using (i) the theorem above repeatedly, we obtain

$$P(A_1 \cap A_2 \cap A_3) = P(A_1) P_{A_1}(A_2) P_{A_1 \cap A_2}(A_3),$$

and generally

$$P(A_1 \cap A_2 \cap \cdots \cap A_n) = P(A_1) P_{A_1}(A_2) \ldots P_{A_1 \cap A_2 \cap \cdots \cap A_{n-1}}(A_n).$$

Let $X(\omega)$ be a random variable on (Ω, P). Setting $A = X^{-1}\{x\}$ $(x \in \Omega^X)$ in $P_A(B)$, we obtain $P_{X^{-1}\{x\}}(B)$. In view of the fact that $X^{-1}\{x\} = \{\omega | X(\omega) = x\}$ we denote $P_{X^{-1}(x)}(B)$ by $P_{X=x}(B)$ or by $P(B|X = x)$ and call it the (*conditional*) *probability* of B under the condition: $X = x$. $P_{X=x}(B)$ is a function of $x \in \Omega^X$ and $B \subset \Omega$. For x fixed $P_{X=x}(B)$ is a probability measure on Ω as a function of B such that $P_{X=x}\{X = x\} = 1$. Hence $P_{X=x}$ is called the (*conditional*) *probability measure* under the condition $X = x$. For B fixed, $P_{X=x}(B)$ is a function of $x \in \Omega^X$. The random variable obtained from this function by replacing x by $X(\omega)$,

$$P_{X=x}(B)|_{x=X(\omega)},$$

is denoted by $P(B|X)$ or $P_X(B)$ and is called the (conditional) probability of B under the condition that the value of $X(\omega)$ is known. Hence P_X is called the (conditional) probability measure under the condition that the value of $X(\omega)$ is known.

Theorem 1.4.2.

$$P(X^{-1}(F) \cap B) = \sum_{x \in F} P\{X = x\} P_{X=x}(B)$$

$$= E(P_X(B), X^{-1}(F)).$$

Proof: Obvious by

$$X^{-1}(F) = \sum_{x \in F} X^{-1}\{x\}$$

and

$$P\{X^{-1}\{x\} \cap B\} = P(X^{-1}\{x\}) P_{X=x}(B). \qquad \blacksquare$$

Since P_A is a probability measure on Ω, we can define the mean value of a real random variable (or random vector) $Y(\omega)$ with respect to P_A, which is denoted by $E_A(Y)$; that is,

$$E_A(Y) = \sum_{\omega \in \Omega} Y(\omega) P_A\{\omega\}.$$

This definition is equivalent to the following:

$$E_A(Y) = \begin{cases} E(Y, A)/P(A), & P(A) > 0, \\ Y(\omega_0), & P(A) = 0 \end{cases}$$

where ω_0 is the fixed point in A used in the definition of $P_A(B)$ in case $P(A) = 0$. Hence it follows that

$$E(Y, A) = P(A) E_A(Y).$$

In the same way as we defined $P_{X=x}(B)$ and $P_X(B)$ we can also define $E_{X=x}(Y)$ and $E_X(Y)$. Corresponding to the theorem above, we immediately obtain

Theorem 1.4.3.

$$E(Y, X^{-1}(F)) = \sum_{x \in F} P\{X = x\} E_{X=x}(Y)$$

$$= E(E_X(Y), X^{-1}(F)).$$

We often denote $E_A(Y)$, $E_{X=x}(Y)$ and $E_X(Y)$ by $E(Y|A)$, $E(Y|X=x)$ and $E(Y|X)$, respectively. For any general random variable $Y(\omega)$, we can define the probability distribution with respect to P_A, which is denoted by $(P_A)^Y$; that is,

$$(P_A)^Y(F) = P_A\{Y^{-1}(F)\}, \qquad F \subset \Omega^Y.$$

Similarly we can define $(P_{X=x})^Y$ and $(P_X)^Y$.

Conditional probabilities are closely related to tree compositions. Let us observe a tree composition

where the probability spaces for T_0 and T_a $(a = a_1, a_2, \ldots, a_n)$ are

$$(\Omega_0 \equiv \{a_1, a_2, \ldots, a_n\}, P_0)$$

and (Ω_a, P_a), respectively. The probability space for the tree composition is given by

$$\Omega = \{(a, b) | a \in \Omega_0, b \in \Omega_a\},$$
$$P\{(a, b)\} = P_0\{a\} P_a\{b\}.$$

The results of the first trial T_0 and the second trial T_a $(a = a_1, a_2, \ldots, a_n)$ are the random variables $X(\omega)$, $Y(\omega)$ on (Ω, P) given by

$$X(\omega) = \pi_1(\omega)$$

and

$$Y(\omega) = \pi_2(\omega) \qquad (\pi_i: i\text{th projection}).$$

It is obvious that

$$\Omega^X = \Omega_0 \qquad \text{and} \qquad \Omega^Y \subset \bigcup_{i=1}^{n} \Omega_{a_i}.$$

Hence we have

$$P\{X = a\} = \sum_{b \in \Omega_a} P\{X = a, Y = b\} = \sum_{b \in \Omega_a} P\{(a, b)\}$$
$$= \sum_{b \in \Omega_a} P_0\{a\} P_a\{b\} = P_0\{a\},$$

and so

$$P_{X=a}\{Y = b\} = \frac{P\{(a, b)\}}{P\{X = a\}} = \frac{P_0\{a\} P_a\{b\}}{P_0\{a\}} = P_a\{b\}.$$

Thus we obtain:

Theorem 1.4.4. $P_{X=a}\{Y = b\} = P_a\{b\}.$

Similarly for a tree composition of several steps. If $X, Y, Z, \ldots$ denote the results of the first step, the second step, the third step, etc., then

$$P_{X=a}\{Y = b\} = P_a\{b\},$$
$$P_{X=a, Y=b}\{Z = c\} = P_{a, b}\{c\},$$
$$\vdots \qquad \vdots$$

For a direct composition, $P_{X=a}\{Y=b\}$ is independent of a, and $P_{X=a,\,Y=b}\{Z=c\}$ is independent of (a,b), and so on.

Exercise 1.4. Prove that if $\Omega = A_1 + A_2 + \cdots + A_n$ ($A_i \neq \emptyset$), then the following equality holds for $B \neq \emptyset$:

$$P_B(A_i) = \frac{P(A_i)P_{A_i}(B)}{\displaystyle\sum_{k=1}^{n} P(A_k)P_{A_k}(B)} \qquad (\textit{Bayes's theorem}).$$

1.5 Independence

Hereafter, whenever we consider several random variables, we assume them to be defined on a probability space unless stated otherwise. Two random variables X and Y are said to be *independent* if we have

$$P\{X=x,\,Y=y\} = P\{X=x\}P\{Y=y\}, \qquad x \in \Omega^X, y \in \Omega^Y.$$

If X and Y are independent, then we have

$$P_{X=x}\{Y=y\} = P\{Y=y\} \ (\text{if } P\{X=x\} > 0), \qquad y \in \Omega^Y,$$

$$P_{Y=y}\{X=x\} = P\{X=x\} \ (\text{if } P\{Y=y\} > 0), \qquad x \in \Omega^X.$$

Conversely each of these conditions implies the independence of X and Y. These can also be stated as follows:

$$P_{X=x}\{Y=y\} \text{ is independent of } x \ (\text{if } P\{X=x\} > 0),$$

$$P_{Y=y}\{X=x\} \text{ is independent of } y \ (\text{if } P\{Y=y\} > 0).$$

The term "independence" is based on this fact.

Theorem 1.5.1. The following conditions are equivalent:
  (i) X and Y are independent.
  (ii) If N denotes the set of all x such that

$$P_{X=x}\{Y=y\} \neq P\{Y=y\},$$

then we have

$$P^X(N) = 0.$$

  (iii) $P_X\{Y=y\} = P\{Y=y\}$ a.s.

Proof: If X and Y are independent, it follows from the observation above that

$$P\{X=x\} = 0 \quad \text{if } P_{X=x}\{Y=y\} \neq P\{Y=y\}.$$

Hence

$$P^X(N) = P\{X \in N\} = \sum_{x \in N} P\{X = x\} = 0.$$

This proves that (i) implies (ii). Set

$$M = \{\omega | P_{X(\omega)}\{Y = y\} \neq P\{Y = y\}\},$$

where $P_{X(\omega)}\{Y = y\}$ denotes $P_{X=\xi}\{Y = y\}|_{\xi = X(\omega)}$, as we explained in the last section. Then $X(\omega) \in N$ for $\omega \in M$, namely, $M \subset X^{-1}(N)$. Hence (ii) implies that

$$P(M) \leqq P(X^{-1}(N)) = P^X(N) = 0; \quad \text{i.e., } P(M) = 0.$$

This proves that (ii) implies (iii). Setting $F = \{x\}$ and $B = Y^{-1}\{y\}$ in Theorem 1.4.2, we have

$$P(X^{-1}\{x\} \cap Y^{-1}\{y\}) = E(P_X(Y^{-1}\{y\}), X^{-1}\{x\}).$$

Hence (iii) implies that

$$P(X^{-1}\{x\} \cap Y^{-1}\{y\}) = E(P(Y^{-1}\{y\}), X^{-1}\{x\})$$
$$= P(Y^{-1}\{y\})P(X^{-1}\{x\}).$$

This proves that (iii) implies (i). ∎

Theorem 1.5.2. If X and Y are independent, then

$$P(X \in E, Y \in F) = P(X \in E)P(Y \in F).$$

Proof: Since

$$\{\omega | X(\omega) \in E, Y(\omega) \in F\} = \sum_{x \in E} \sum_{y \in F} \{\omega | X(\omega) = x, Y(\omega) = y\},$$

we have

$$P\{X \in E, Y \in F\} = \sum_{x \in E} \sum_{y \in F} P\{X = x\}P\{Y = y\}$$
$$= P(X \in E)P(Y \in F).$$

∎

Independence of more than two random variables is defined similarly: The system $\{X_1, X_2, \ldots, X_n\}$ is said to be independent if

$$P\{X_1 = x_1, X_2 = x_2, \ldots, X_n = x_n\}$$
$$= P\{X_1 = x_1\}P\{X_2 = x_2\}\ldots P\{X_n = x_n\}, \quad x_i \in \Omega^{X_i}, i = 1, 2, \ldots, n.$$

It is obvious that independence is preserved by any permutation of random variables. Also, any subsystem of an independent system is independent, as we can see by adding each side of the equality above.

Theorem 1.5.3. Decompose $\{X_1, X_2, \ldots, X_n\}$ into two subsystems $\{X_1, X_2, \ldots, X_r\}$ and $\{X_{r+1}, X_{r+2}, \ldots, X_n\}$. $\{X_1, X_2, \ldots, X_n\}$ is independent if and only if the following two conditions are satisfied:

  (i) $\{X_1, X_2, \ldots, X_r\}$ is independent and $\{X_{r+1}, X_{r+2}, \ldots, X_n\}$ is independent.
  (ii) The joint variables $(X_1, X_2, \ldots, X_r)$ and $(X_{r+1}, \ldots, X_n)$ are independent.

Similarly for a decomposition into more than two subsystems.

Proof: We can easily prove this theorem, noting that
$$X_1 = x_1, \ldots, X_r = x_r, X_{r+1} = x_{r+1}, \ldots, X_n = x_n$$
$$\Leftrightarrow (X_1, \ldots, X_r) = (x_1, \ldots, x_r), (X_{r+1}, \ldots, X_n) = (x_{r+1}, \ldots, x_n).$$
∎

Theorem 1.5.4. If $\{X_1, X_2, \ldots, X_n\}$ is independent, then
$$P\{X_1 \in E_1, X_2 \in E_2, \ldots, X_n \in E_n\}$$
$$= P\{X_1 \in E_1\} P\{X_2 \in E_2\} \ldots P\{X_n \in E_n\}.$$

Proof: Using the theorems above, we can easily see that the left-hand side
$$= P\{(X_1, X_2, \ldots, X_{n-1}) \in E_1 \times E_2 \times \cdots \times E_{n-1}, X_n \in E_n\}$$
$$= P\{(X_1, X_2, \ldots, X_{n-1}) \in E_1 \times E_2 \times \cdots \times E_{n-1}\} P\{X_n \in E_n\}$$
$$= P\{X_1 \in E_1, X_2 \in E_2, \ldots, X_{n-1} \in E_{n-1}\} P\{X_n \in E_n\}.$$

Now use the induction to complete the proof. ∎

This theorem means that if $\{X_1, X_2, \ldots, X_n\}$ is independent, then
$$P^{(X_1, X_2, \ldots, X_n)}(E_1 \times E_2 \times \cdots \times E_n)$$
$$= P^{X_1}(E_1) P^{X_2}(E_2) \ldots P^{X_n}(E_n);$$
that is,
$$P^{(X_1, X_2, \ldots, X_n)} = P^{X_1} \times P^{X_2} \times \cdots \times P^{X_n}.$$

Theorem 1.5.5. Suppose that $Y_1 = \varphi_1(X_1)$, $Y_2 = \varphi_2(X_2), \ldots, Y_n = \varphi_n(X_n)$. Then independence of $\{X_1, X_2, \ldots, X_n\}$ implies that of $\{Y_1, Y_2, \ldots, Y_n\}$.

Proof: Use the last theorem, noting that
$$Y_1 = y_1, Y_2 = y_2, \ldots, Y_n = y_n \Leftrightarrow X_1 \in \varphi_1^{-1}(y_1), X_2 \in \varphi_2^{-1}(y_2), \ldots,$$
$$X_n \in \varphi_n^{-1}(y_n).$$
∎

A system of sets $\{A_1, A_2, \ldots, A_n\}$ is said to be independent if the system of their indicators $1_{A_1}(\omega), 1_{A_2}(\omega), \ldots, 1_{A_n}(\omega)$ is independent. Since a set represents an event, this defines the independence of a system of events.

Theorem 1.5.6. The following three conditions are equivalent:
  (i) $\{A_1, A_2, \ldots, A_n\}$ is independent.
  (ii) $P(A'_1 \cap A'_2 \cap \cdots \cap A'_n) = P(A'_1)P(A'_2)\ldots P(A'_n)$, where $A'_i = A_i$ or A^c_i, $i = 1, 2, \ldots, n$.
  (iii) $P(A_{i_1} \cap A_{i_2} \cap \cdots \cap A_{i_k}) = P(A_{i_1})P(A_{i_2})\ldots P(A_{i_k})$ for every $k = 2, 3, \ldots, n$ and every $1 \leq i_1 < i_2 < \cdots < i_k \leq n$.

Proof: Let $e_i(\omega)$ denote the indicator of A_i. Then (i) is equivalent to

$$P\big(e_1^{-1}(a_1) \cap e_2^{-1}(a_2) \cap \cdots \cap e_n^{-1}(a_n)\big)$$
$$= P\big(e_1^{-1}(a_1)\big)P\big(e_2^{-1}(a_2)\big)\ldots P\big(e_n^{-1}(a_n)\big)$$

for $a_i = 0, 1$ $(i = 1, 2, \ldots, n)$. Hence (i) and (ii) are equivalent to each other because $e_i^{-1}(1) = A_i$ and $e_i^{-1}(0) = A^c_i$. Since any subsystem of an independent system is independent, (ii) implies (iii). It remains only to prove that (iii) implies (ii). If

$$\sum_{k, \{i_\nu\}} c_{i_1 i_2 \ldots i_k} t_{i_1} t_{i_2} \ldots t_{i_k}$$

is the expansion of

$$(a_1 + b_1 t_1)(a_2 + b_2 t_2)\ldots(a_n + b_n t_n),$$

then we have

$$E\big((a_1 + b_1 e_1)(a_2 + b_2 e_2)\ldots(a_n + b_n e_n)\big)$$
$$= \sum_{k, \{i_\nu\}} c_{i_1 i_2 \ldots i_k} E\big(e_{i_1} e_{i_2} \ldots e_{i_k}\big),$$
$$E(a_1 + b_1 e_1)E(a_2 + b_2 e_2)\ldots E(a_n + b_n e_n)$$
$$= \sum_{k, \{i_\nu\}} c_{i_1 i_2 \ldots i_k} E(e_{i_1})E(e_{i_2})\ldots E(e_{i_k}).$$

Suppose that (iii) holds. Then

$$E\big(e_{i_1} e_{i_2} \ldots e_{i_k}\big) = E(e_{i_1})E(e_{i_2})\ldots E(e_{i_k})$$

because $E(1_A) = P(A)$. Hence

$$E\big((a_1 + b_1 e_1)(a_2 + b_2 e_2)\ldots(a_n + b_n e_n)\big)$$
$$= E(a_1 + b_1 e_1)E(a_2 + b_2 e_2)\ldots E(a_n + b_n e_n).$$

Let e_i' denote the indicator of A_i'. Since $A_i' = A_i$ or A_i^c, we have $e_i' = e_i$ or $1 - e_i$. Hence e_i' is of the form $a + be_i$. Therefore

$$E(e_1'e_2'...e_n') = E(e_1')E(e_2')...E(e_n');$$

that is,

$$P(A_1' \cap A_2' \cap \cdots \cap A_n') = P(A_1')P(A_2')...P(A_n').$$

This proves that (iii) implies (ii). ∎

Let $\mathbf{T}$ be the direct composition of $\mathbf{T}_1, \mathbf{T}_2, \ldots, \mathbf{T}_n$ and let (Ω_1, P_1), $(\Omega_2, P_2), \ldots, (\Omega_n, P_n)$, and (Ω, P) be the probability spaces for $\mathbf{T}_1, \mathbf{T}_2, \ldots, \mathbf{T}_n$, and $\mathbf{T}$, respectively. Then

$$\Omega = \Omega_1 \times \Omega_2 \times \cdots \times \Omega_n, \qquad P = P_1 \times P_2 \times \cdots \times P_n.$$

If ω_i is an outcome of $\mathbf{T}_i$ for $i = 1, 2, \ldots, n$, then $\omega = (\omega_1, \omega_2, \ldots, \omega_n)$ is an outcome of $\mathbf{T}$. Hence ω_i is represented as

$$\omega_i = \pi_i(\omega) \qquad (\pi_i: i\text{-projection})$$

on (Ω, P), and a random variable $X_i(\omega_i)$ on (Ω_i, P_i) is represented as

$$X_i(\omega_i) = X_i(\pi_i(\omega)) = (X_i \circ \pi_i)(\omega)$$

on (Ω, P).

Theorem 1.5.7. $\{\omega_1, \omega_2, \ldots, \omega_n\}$ is independent on (Ω, P) and $\{X_1, X_2, \ldots, X_n\}$ is independent.

Proof:

$$P\{\pi_1(\omega) = a_1, \pi_2(\omega) = a_2, \ldots, \pi_n(\omega) = a_n\}$$
$$= P\{(a_1, a_2, \ldots, a_n)\} = P_1\{a_1\}P_2\{a_2\}...P_n\{a_n\},$$
$$P\{\pi_1(\omega) = a_1\} = P(\{a_1\} \times \Omega_2 \times \Omega_3 \times \cdots \times \Omega_n)$$
$$= P_1\{a_1\}P_2(\Omega_2)P_3(\Omega_3)...P_n(\Omega_n)$$
$$= P_1\{a_1\}.$$

Similarly, we have

$$P\{\pi_i(\omega) = a_i\} = P_i\{a_i\}$$

for every i. Hence

$$P\{\pi_1(\omega) = a_1, \pi_2(\omega) = a_2, \ldots, \pi_n(\omega) = a_n\}$$
$$= P\{\pi_1(\omega) = a_1\}P\{\pi_2(\omega) = a_2\}...P\{\pi_n(\omega) = a_n\}.$$

This implies the independence of $\omega_1 = \pi_1(\omega)$, $\omega_2 = \pi_2(\omega), \ldots, \omega_n = \pi_n(\omega)$. Hence $\{X_1(\omega_1), X_2(\omega_2), \ldots, X_n(\omega_n)\}$ is also independent, by Theorem 1.5.5. ∎

Remark: In view of this theorem a direct composition is also called an *independent composition*.

Exercise 1.5. Prove the following statements.

(i) Let X be the number obtained when throwing a die and A_i the event that X is a multiple of i. Then A_2 and A_3 are independent. [*Hint*: By Theorem 1.5.6 (i)–(iii), it suffices to show that $P(A_2 \cap A_3) = P(A_2)P(A_3)$.]

(ii) Let $X_1, X_2, \ldots, X_{m+n}$ be the numbers obtained when throwing a die $(m + n)$ times and let $S = X_1 + X_2 + \cdots + X_m$ and $T = X_{m+1} + X_{m+2} + \cdots + X_{m+n}$. Then S and T are independent. [*Hint*: Use Theorems 1.5.3 and 1.5.5.]

(iii) If (a) X_1 and X_2 are independent, (b) $\{X_1, X_2\}$ and X_3 are independent, and (iii) $\{X_1, X_2, X_3\}$ and X_4 are independent, then $\{X_1, X_2, X_3, X_4\}$ is independent. The converse is also true. [*Hint*: Use Theorem 1.5.3.]

(iv) If A and A are independent, then $P(A) = 0$ or 1. [*Hint*: $A \cap A = A$.]

(v) If X and X are independent, then $X = a$ a.s. for some $a \in \Omega^X$. [*Hint*: Use (iv) to prove that $P\{X = a\} = 1$ or 0 for every $a \in \Omega^X$.]

1.6 Independent random variables

In this section $X, X_1, X_2, \ldots$ are real random variables.

Theorem 1.6.1 (Multiplicativity of mean values). If $\{X_1, X_2, \ldots, X_n\}$ is independent, then

$$E(X_1 X_2 \ldots X_n) = EX_1 EX_2 \ldots EX_n.$$

Proof: The case $n = 1$ is trivial. Consider the case $n = 2$. By Theorem 1.2.2(v) we have

$$E(X_1 X_2) = \sum x_1 x_2 P^{(X_1, X_2)}\{(x_1, x_2)\},$$

where $\sum$ is the sum for x_1 and x_2 running over Ω^{X_1} and Ω^{X_2}, respectively. Since X_1 and X_2 are independent, we have

$$P^{(X_1, X_2)}\{(x_1, x_2)\} = P^{X_1}\{x_1\} P^{X_2}\{x_2\}.$$

Observing that

$$EX_i = \sum x_i P^{X_i}\{x_i\}$$

we obtain

$$E(X_1 X_2) = EX_1 EX_2.$$

Since $\{X_1, X_2, \ldots, X_n\}$ is independent, the product $X_1 \cdot X_2 \cdot X_3 \ldots X_{n-1}$ and X_n are independent by Theorems 1.5.3 and 1.5.5. Using the case $n = 2$ proved above, we obtain

$$E(X_1 X_2 \ldots X_{n-1} X_n) = E(X_1 X_2 \ldots X_{n-1}) EX_n.$$

Now use induction to complete the proof. ∎

The additivity of mean values

$$E(X_1 + X_2 + \cdots + X_n) = EX_1 + EX_2 + \cdots + EX_n$$

holds for any real random variables, but the multiplicativity of mean values holds only for independent random variables. In fact, if $X_1 = X_2 = X$, then

$$E(X_1 X_2) - EX_1 EX_2 = EX^2 - (EX)^2 = V(X) \geq 0,$$

and the equality does not hold in general.

Theorem 1.6.2 (Additivity of variances). If any two of $X_1, X_2, \ldots, X_n$ are independent, then

$$V(X_1 + X_2 + \cdots + X_n) = V(X_1) + V(X_2) + \cdots + V(X_n).$$

Proof: Set $Y_i = X_i - EX_i$. Then Y_i and Y_j are independent for $i \neq j$ and

$$V(X_i) = E(Y_i^2),$$
$$V(X_1 + X_2 + \cdots + X_n) = E\left((Y_1 + Y_2 + \cdots + Y_n)^2\right).$$

By multiplicativity of mean values we have

$$E(Y_i Y_j) = EY_i EY_j = 0, \qquad i \neq j.$$

Hence

$$E\left((Y_1 + Y_2 + \cdots + Y_n)^2\right) = \sum_i E(Y_i^2) + 2\sum_{i<j} E(Y_i Y_j) = \sum_i E(Y_i^2),$$

namely,

$$V(X_1 + X_2 + \cdots + X_n) = V(X_1) + V(X_2) + \cdots + V(X_n). \quad ∎$$

If $\{X_1, X_2, \ldots, X_n\}$ is independent, then the assumption of this theorem holds, so the additivity of variances follows. Note that the additivity of variances does not hold for a general system of random variables. For example, if $X_1 = X_2 = \cdots = X_n = X$ then

$$V(X_1 + X_2 + \cdots + X_n) = n^2 V(X),$$
$$V(X_1) + V(X_2) + \cdots + V(X_n) = nV(X).$$

If X is a real random variable, then so is t^X for $t \in (0,1)$. The function of t

$$g^X(t) = E(t^X), \qquad t \in (0,1),$$

is called the *generating function* of X. By Theorem 1.2.2(v) we have

$$g^X(t) = \sum_{x \in \Omega^X} t^x P^X\{x\}.$$

That is, $P^X\{x\}$ is the coefficient of t^x in the expansion of $g^X(t)$ in t, which is determined by the following recursion formula:

$$P^X\{x_1\} = \lim_{t \downarrow 0} t^{-x_1} g^X(t),$$

$$P^X\{x_k\} = \lim_{t \downarrow 0} t^{-x_k} \left(g^X(t) - \sum_{i=1}^{k-1} t^{x_i} P^X\{x_i\} \right),$$

where $x_1 < x_2 < \cdots < x_n$ are the points in $\Omega^X \subset \mathbb{R}$.

If $X_1, X_2, \ldots, X_n$ are independent, then so are $t^{X_1}, t^{X_2}, \ldots, t^{X_n}$. Using the multiplicativity of mean values, we obtain the following:

Theorem 1.6.3 (Multiplicativity of generating functions). Let $X_1, X_2, \ldots, X_n$ be independent and $X = X_1 + X_2 + \cdots + X_n$. Then

$$g^X(t) = g^{X_1}(t) g^{X_2}(t) \ldots g^{X_n}(t).$$

We will show two elementary examples of application of this theorem.

Let $A_1, A_2, \ldots, A_n$ be independent events with the identical probability p, and let N denote the number of the occurring events among these events. Then we have

$$P(N = k) = \binom{n}{k} p^k (1-p)^{n-k}, \qquad k = 0, 1, 2, \ldots, n.$$

To prove this, denote the indicator of A_i by $e_i(\omega)$. Then

$$g^{e_i}(t) = t^1 P(e_i = 1) + t^0 P(e_i = 0) = tP(A_i) + P(A_i^c) = tp + (1-p).$$

N is a random variable represented as

$$N = e_1 + e_2 + \cdots + e_n.$$

Since $\{e_1, e_2, \ldots, e_n\}$ is independent because of the independence of $\{A_1, A_2, \ldots, A_n\}$, we obtain

$$g^N(t) = \prod_{i=1}^{n} g^{e_i}(t) = (tp + (1-p))^n = \sum_{k=0}^{n} \binom{n}{k} p^k (1-p)^{n-k} t^k,$$

so

$$P(N = k) = P^N\{k\} = \binom{n}{k} p^k (1-p)^{n-k}.$$

Let S be the sum of the numbers obtained when throwing a die three times. We will find $P_k = P(S = k)$. Denote by X_i the number obtained in the ith throw. Then

$$g^{X_i}(t) = \tfrac{1}{6}(t + t^2 + \cdots + t^6) = \tfrac{1}{6}(1-t)^{-1}(1-t^6)t.$$

Since $\{X_1, X_2, X_3\}$ is independent and $S = X_1 + X_2 + X_3$, we can use the multiplicativity of generating functions to obtain

$$g^S(t) = \prod_{i=1}^{3} g^{X_i}(t) = 6^{-3}(1-t)^{-3}(1-t^6)^3 t^3$$

$$= 6^{-3}\left(1 + 3t + \frac{3\cdot 4}{2}t^2 + \frac{4\cdot 5}{2}t^3 + \frac{5\cdot 6}{2}t^4 + \cdots\right)$$

$$\times (1 - 3t^6 + 3t^{12} - t^{18})t^3.$$

Thus P_k equals the coefficients of t^k in the expansion of this function. For example,

$$P_9 = 6^{-3}\left(\frac{7\cdot 8}{2} - 3\right) = \frac{25}{216},$$

$$P_{10} = 6^{-3}\left(\frac{8\cdot 9}{2} - 9\right) = \frac{27}{216}.$$

Exercise 1.6. Use the multiplicativity of mean values to prove that if $\{X_1, X_2, \ldots, X_n\}$ is independent and if $EX_i = 0$ and $EX_i^2 = 1$ for every i, then

$$E\left((X_1 + X_1 X_2 + X_1 X_2 X_3 + \cdots + X_1 X_2 X_3 \ldots X_n)^2\right) = n.$$

1.7   The law of large numbers

When we throw a die many times, the ratio of the number of the throws yielding i ($i = 1, 2, \ldots, 6$) to the total number of throws is nearly equal to $\tfrac{1}{6}$. This fact is familiar to everybody, and is known as *the law of large numbers.* The aim of this section is to formulate this experimental fact mathematically and to prove it rigorously.

The probability space of the trial of throwing a die n times is given by

$$\Omega = \Omega_0^n, \qquad P = P_0^n,$$

where

$$\Omega_0 = \{1, 2, \ldots, 6\},$$

$$P_0\{i\} = \tfrac{1}{6}, \qquad i = 1, 2, \ldots, 6.$$

Let $X_k = X_k(\omega)$ be the number obtained in the kth throw for $k = 1, 2, \ldots, n$. Then X_k is represented on (Ω, P) as follows:

$$X_k(\omega) = \pi_k(\omega) \qquad (\pi_k\colon \text{the } k\text{-projection}; \; k = 1, 2, \ldots, n).$$

It is obvious that $X_1, X_2, \ldots, X_n$ are independent random variables with the identical probability distribution

$$P\{X_k(\omega) = i\} = \tfrac{1}{6}, \qquad k = 1, 2, \ldots, n.$$

Using the map

$$1_i\colon \Omega_0 \to \{0, 1\}, \qquad 1_i(i) = 1, \qquad 1_i(j) = 0 \qquad (j \neq i),$$

we can express the frequency of the number of throws yielding i, $R_i = R_i(\omega)$, as follows:

$$R_i = R_i(\omega) = \frac{1}{n} \sum_{k=1}^{n} 1_i(X_k(\omega)).$$

Hence R_i is a random variable on (Ω, P). Although both R_i and (Ω, P) depend on n, we omit n for simplicity of notation. The law of large numbers claims that

$$R_i(\omega) \doteqdot \tfrac{1}{6}, \qquad i = 1, 2, \ldots, 6$$

for large n.

Now we will check the meaning of this approximate expression. One may interpret it in the sense that for every $\varepsilon > 0$, we can find $n_0(\varepsilon)$ such that

$$\max_{\omega} |R_i(\omega) - \tfrac{1}{6}| < \varepsilon \qquad \text{for } n > n_0(\varepsilon).$$

This is obviously absurd because $R_i(\omega)$ may take 1, however large n is. But if n is large, the probability that $R_i(\omega) = 1$ is very small. Hence we will interpret the law of large numbers probabilistically as follows:

Theorem 1.7 (Bernoulli's law of large numbers). For every $\varepsilon > 0$ we can find $n_0(\varepsilon)$ such that

$$P\left\{\left|R_i - \tfrac{1}{6}\right| > \varepsilon\right\} < \varepsilon, \qquad i = 1, 2, \dots, 6.$$

for $n > n_0(\varepsilon)$.

Proof: Fixing i, we set $e_k(\omega) = 1_i(X_k(\omega))$. The independence of $\{X_1, X_2, \dots, X_n\}$ implies that of $\{e_1, e_2, \dots, e_n\}$. By additivity of mean values and variances we obtain

$$E(e_k) = 1P\{X_k = i\} + 0P\{X_k \neq i\} = \tfrac{1}{6},$$

$$E(R_i) = E\left(\frac{e_1 + e_2 + \cdots + e_n}{n}\right) = \frac{E(e_1) + E(e_2) + \cdots + E(e_n)}{n} = \frac{1}{6},$$

$$V(e_i) = E\left((e_i - \tfrac{1}{6})^2\right) = \left(\tfrac{5}{6}\right)^2 P\{X_k = i\} + \left(\tfrac{1}{6}\right)^2 P\{X_k \neq i\} = \tfrac{5}{36},$$

$$V(R_i) = V\left(\frac{e_1 + e_2 + \cdots + e_n}{n}\right) = \frac{1}{n^2} V(e_1 + e_2 + \cdots + e_n)$$

$$= \frac{1}{n^2}\left(V(e_1) + V(e_2) + \cdots + V(e_n)\right) = \frac{5}{36n},$$

$$\sigma(R_i) - \frac{\sqrt{5}}{6\sqrt{n}}.$$

Using Chebyshev's inequality (Theorem 1.2.7), we get

$$P\left\{\left|R_i - \frac{1}{6}\right| > a\frac{\sqrt{5}}{6\sqrt{n}}\right\} \leq \frac{1}{a^2}.$$

Take a large number $a = a(\varepsilon)$ that makes the right-hand side less than ε, and then take $n_0 = n_0(\varepsilon)$ for this $a(\varepsilon)$ so that

$$a\frac{\sqrt{5}}{6\sqrt{n_0}} < \varepsilon.$$

Then

$$P\left\{\left|R_i - \frac{1}{6}\right| > \varepsilon\right\} < \varepsilon \qquad \text{for } n > n_0(\varepsilon). \qquad \blacksquare$$

Exercise 1.7. Formulate the law of large numbers concerning the repetition of a general trial and prove it.

2

Probability measures

2.1   General trials and probability measures

In Chapter 1 we explained the fundamental idea of probability theory concerning finite trials. From now on we will discuss those trials that may be infinite in general. The way of developing the theory is essentially the same as in the case of finite trials. However, there are several situations which need more sophisticated consideration.

Let T be a general trial, Ω the sample space, and P the probability law. In the case of finite trials $P(A)$ was defined for every subset A of Ω, namely, $\mathscr{D}(P) = 2^\Omega$. In the case of infinite trials we assume that $\mathscr{D}(P)$ is a subclass of 2^Ω satisfying the following three conditions:

(σ.1) $\Omega \in \mathscr{D}(P)$;

(σ.2) $A \in \mathscr{D}(P) \Rightarrow A^c \in \mathscr{D}(P)$;

(σ.3) $A_n \in \mathscr{D}(P)$ $(n = 1, 2, \ldots) \Rightarrow \bigcup_{n=1}^\infty A_n \in \mathscr{D}(P)$.

The first condition is a natural requirement. The second claims that if the probability of an event is defined, then the probability of its complementary event should be defined. The third claims that if the probability of A_n is defined for every n, then the probability of occurrence of at least one of $A_1, A_2, \ldots$ should be defined. The second and third conditions are also natural. In general a class of subsets of Ω satisfying these three conditions is called a σ-*algebra*, one of the most fundamental concepts in measure theory.

Next we impose the following three assumptions on the value $P(A)$, where $A \in \mathscr{D}(P)$ is implicitly assumed whenever we write $P(A)$.

(P.1) $P(A) \geqq 0$;

(P.2) $P(\sum_{n=1}^\infty A_n) = \sum_{n=1}^\infty P(A_n)$ (σ-additivity);

(P.3) $P(\Omega) = 1$.

The first and third conditions are the same as in the case of finite trials. The second condition, "σ-additivity," claims that if $A_1, A_2, \ldots$ are exclusive events, then the probability of occurrence of at least one of $A_1, A_2, \ldots$ equals the sum of the probabilities $P(A_n)$, $n = 1, 2, \ldots$. In general this condition is stronger than the additivity that was imposed in the case of finite trials. σ-additivity holds true and plays a very important role in practically all useful cases. Needless to say, σ-additivity and additivity are equivalent to each other for the case of finite trials.

A set function satisfying (P.1) and (P.2) is called a *measure*, and if it further satisfies (P.3), it is called a *probability measure*. A set A is called *P-measurable* if it belongs to $\mathscr{D}(P)$, and $P(A)$ is called the *P-measure* (or simply measure) of A. A set Ω endowed with a probability measure P on Ω is called a *probability space* (Ω, P).

Let (Ω, P) be a general probability space. Then we can imagine a trial of drawing a point $\omega \in \Omega$ such that a point of $A \in \mathscr{D}(P)$ be drawn with probability $P(A)$. Such a trial is called a *sampling* from (Ω, P). The sample space and the probability measure for this trial are obviously Ω and P, respectively. A trial T whose sample space and probability measure are respectively, Ω and P is essentially the same as a sampling from the probability space (Ω, P).

Let Ω be a unit interval $[0,1]$ and P the Lebesgue measure on $[0,1]$, where $\mathscr{D}(P)$ is the class of all Lebesgue-measurable sets. A sampling from (Ω, P) is called a *random sampling* from $[0,1]$. If we take, instead of the Lebesgue measure, the probability measure

$$P(E) = \int_E f(\omega)\, d\omega,$$

$f(x)$ being a Lebesgue-measurable function satisfying

$$f(\omega) \geqq 0, \qquad \int_{[0,1]} f(\omega)\, d\omega = 1,$$

then a sampling from (Ω, P) is called a random sampling from $[0,1]$ with *weight* $f(\omega)$. The function f satisfying the conditions above is called a *probability density* on $[0,1]$ and the probability measure on $[0,1]$ defined above is called a probability measure with density f.

Not every probability measure on $\Omega = [0,1]$ has density. For example, fix a point a in $\Omega = [0,1]$ and define $P(E)$ as follows:

$$P(E) = \delta_a(E) = \begin{cases} 1, & a \in E, \\ 0, & a \notin E. \end{cases}$$

Then P is also a probability measure on Ω with $\mathscr{D}(P) = 2^{\Omega}$. A sampling from (Ω, δ_a) means drawing a almost surely.

Let us give a more complicated sampling from $[0,1]$. Arrange all rational numbers in $\Omega = [0,1]$ as $r_1, r_2, \ldots$ and define P by

$$P(E) = \sum_{i:\, r_i \in E} 2^{-i}.$$

It is easy to check that P is a probability measure on Ω with $\mathscr{D}(P) = 2^{\Omega}$. Denoting $\{r_1, r_2, \dots\}$ by $\mathbf{Q}_{[0,1]}$, we have

$$P(\mathbf{Q}_{[0,1]}) = 1, \qquad P(\Omega - \mathbf{Q}_{[0,1]}) = 0,$$

so that only rational numbers appear almost surely in a sampling from (Ω, P).

Let us consider a trial of tossing a coin infinitely many times. Denoting heads by 1 and tails by 0, we can represent a sample point of this trial by an infinite sequence with terms 0 or 1. For example, $(0,0,1,0,1,\dots)$ means that tails, tails, heads, tails, heads,... appear in this order. Hence the sample space Ω is as follows:

$$\Omega = \{\omega = (\omega_1, \omega_2, \dots) | \omega_n = 0,1 (n = 1,2,\dots)\}.$$

In the first throw heads and tails appear with probability $\frac{1}{2}$. Since these events are represented on Ω, respectively, by the following sets:

$$\Omega_1 = \{\omega = (1, \omega_2, \omega_3, \dots) | \omega_n = 0,1 \ (n = 2,3,\dots)\},$$
$$\Omega_0 = \{\omega = (0, \omega_2, \omega_2, \dots) | \omega_n = 0,1 \ (n = 2,3,\dots)\},$$

we have

$$P(\Omega_0) = P(\Omega_1) = \tfrac{1}{2}.$$

Similarly, in the first and second throws heads-heads, heads-tails, tails-heads, and tails-tails appear with probability $\frac{1}{4}$. Hence, setting

$$\Omega_{ij} = \{\omega = (i, j, \omega_3, \omega_4, \dots) | \omega_n = 0,1 \ (n = 3,4,\dots)\}, \qquad i, j = 0,1,$$

we have

$$P(\Omega_{00}) = P(\Omega_{01}) = P(\Omega_{10}) = P(\Omega_{11}) = \tfrac{1}{4}.$$

In general, setting

$$\Omega_{i_1 i_2 \dots i_k} = \{\omega = (i_1, i_2, \dots, i_k, \omega_{k+1}, \omega_{k+2}, \dots) | \omega_n = 0,1 \ (n \geq k+1)\},$$

$$i_1, i_2, \dots, i_k = 0,1$$

we have

$$P(\Omega_{i_1 i_2 \dots i_k}) = 2^{-k}.$$

Thus we have defined $P(I)$ for every I belonging to the class of subsets of Ω:

$$\mathscr{I} = \{\Omega_{i_1 i_2 \dots i_k} | i_\nu = 0,1 \ (\nu = 1,2,\dots,k), k = 1,2,\dots\}.$$

If we can extend this set function to a probability measure on Ω, then the extension is the probability measure for our trial. To do this we need the extension theorem that will be discussed in the next section; see Exercise 2.2.

We will review some fundamental facts in measure theory in the exercise below. These facts are frequently used throughout this book.

Exercise 2.1.

(i) Prove that the following set functions $N_{m,v}$ and $C_{m,c}$ defined for all Lebesgue-measurable sets are probability measurers on $\mathbb{R}^1$, where $m \in \mathbb{R}^1$, $v > 0$, and $c > 0$.

$$N_{m,v}(E) = \int_E \frac{1}{\sqrt{2\pi v}} e^{-(x-m)^2/2v} \, dx \qquad (Gauss\ distribution),$$

$$C_{m,c}(E) = \int_E \frac{c}{\pi} \frac{1}{c^2 + (x-m)^2} \, dx \qquad (Cauchy\ distribution).$$

(ii) Prove that every σ-algebra $\mathscr{B}$ on a set S contains ϕ and S and is closed under countable unions, countable intersections, complements, differences, upper limits, and lower limits. (A class $\mathscr{C}$ of subsets of a set S is said to be *closed* under a set operation α if α carries $\mathscr{C}$ into itself.) [*Hint*: $\mathscr{B}$ contains S and is closed under countable infinite unions and complements by the definition of a σ-algebra. Hence $\phi = S^c \in \mathscr{B}$. If $A, B \in \mathscr{B}$, then

$$A \cup B = A \cup B \cup \varnothing \cup \varnothing \cup \ldots \in \mathscr{B}$$

Thus $\mathscr{B}$ is closed under finite unions. Note the following equalities for the other operations:

$$\bigcap_n A_n = \left(\bigcup_n A_n^c\right)^c, \qquad A \backslash B = A \cap B^c,$$

$$\limsup_{n\to\infty} A_n = \bigcap_k \bigcup_{n>k} A_n, \qquad \liminf_{n\to\infty} A_n = \bigcup_k \bigcap_{n>k} A_n.]$$

(iii) Prove that every probability measure P has the following properties:

$$P(\phi) = 0, \qquad P(A^c) = 1 - P(A), \qquad 0 \leq P(A) \leq 1,$$

$$P(A + B) = P(A) + P(B) \qquad (finite\ additivity),$$

$$A \subset B \Rightarrow P(A) \leq P(B) \qquad (monotonicity),$$

$$P(B - A) = P(B) - P(A),$$

$$\{A_n\} \text{ monotone} \Rightarrow P\left(\lim_{n\to\infty} A_n\right) = \lim_{n\to\infty} P(A_n)$$

(monotone convergence theorem),

$$P\left(\liminf_{n\to\infty} A_n\right) \leq \liminf_{n\to\infty} P(A_n) \leq \limsup_{n\to\infty} P(A_n)$$

$$\leq P\left(\limsup_{n\to\infty} A_n\right) \quad \text{(Fatou's lemma)},$$

$$A = \lim_{n\to\infty} A_n \Rightarrow P(A) = \lim_{n\to\infty} P(A_n) \quad \text{(continuity)},$$

$$P\left(\bigcup_n A_n\right) \leq \sum_n P(A_n) \quad \text{(subadditivity)},$$

$$P\left(\bigcap_n A_n\right) \geq 1 - \sum_n P(A_n^c),$$

$$P(A_n) = 0 \ (n = 1, 2, \ldots) \Rightarrow P\left(\bigcup_n A_n\right) = 0,$$

$$P(A_n) = 1 \ (n = 1, 2, \ldots) \Rightarrow P\left(\bigcap_n A_n\right) = 1.$$

[*Hint*: $\Omega = \Omega + \varnothing + \varnothing + \cdots$ and the σ-additivity of P imply that $1 = 1 + P(\varnothing) + P(\varnothing) + \cdots$, so $P(\varnothing) = 0$. Hence we have the finite additivity of P:

$$P(A + B) = P(A + B + \varnothing + \varnothing + \cdots)$$

$$= P(A) + P(B) + 0 + 0 + \cdots$$

$$= P(A) + P(B).$$

This implies the properties on the first four lines. To prove the monotone convergence theorem for an increasing sequence of sets, note that

$$A_1 \subset A_2 \subset \cdots \Rightarrow \lim_{n\to\infty} A_n = A_1 + (A_2 - A_1) + (A_3 - A_2) + \cdots$$

and reduce the case of a decreasing sequence to this case by considering complements. From this theorem follows Fatou's lemma, which implies the property of continuity. Note that

$$\bigcup_n A_n = \sum_n B_n, \quad B_n = A_n \setminus \bigcup_{i=1}^{n-1} A_i \subset A_n$$

to prove the subadditivity. The other properties will follow easily.]

(iv) Prove that if P is a probability measure, then

$$P\left(\bigcup_n A_n \setminus \bigcup_n B_n\right) \le \sum_n P(A_n \setminus B_n),$$

$$P\left(\bigcap_n A_n \setminus \bigcap_n B_n\right) \le \sum_n P(A_n \setminus B_n).$$

[*Hint*: Note that

$$\bigcup_n A_n \setminus B = \bigcup_n (A_n \setminus B) \subset \bigcup_n (A_n \setminus B_n), \qquad \text{where } B = \bigcup_n B_n,$$

to prove the first inequality, to which the second one is reduced by observing that $A \setminus B = B^c \setminus A^c$.

(v) Prove that $x \in \limsup A_n$ if and only if x is contained in an infinite number of the sets $A_1, A_2, \ldots$; and that $x \in \liminf A_n$ if and only if x is contained in all the sets $A_1, A_2, \ldots$ but a finite number of exceptions.

(vi) Prove that for every $\mathscr{A} \subset 2^S$ there exists the smallest σ-algebra, $\sigma[\mathscr{A}]$, on S including $\mathscr{A}$, which is called the σ-algebra generated by $\mathscr{A}$. [*Hint*: $\sigma[\mathscr{A}]$ is the intersection of all σ-algebras on S containing $\mathscr{A}$.]

(vii) Let S be a topological space. The σ-algebra on S generated by the open subsets of S is called the *Borel system* on S, written $\mathscr{B}(S)$. An element of $\mathscr{B}(S)$ is called a *Borel subset* of S. Prove that every interval, open, closed or semiopen, in $\mathbb{R}^1$ is a Borel subset of $\mathbb{R}^1$ and that $\mathscr{B}(\mathbb{R}^1)$ is generated by the open intervals of $\mathbb{R}^1$.

(viii) Let $\mathscr{B}_i$ be a σ-algebra on S_i for $i = 1, 2, \ldots, n$, and S be the Cartesian product $S_1 \times S_2 \times \cdots \times S_n$. The σ-algebra on S generated by all subsets of the form

$$B_1 \times B_2 \times \cdots \times B_n, \qquad B_i \in B_i, i = 1, 2, \ldots, n,$$

is called the *product σ-algebra* of $\mathscr{B}_i$, $i = 1, 2, \ldots, n$, written

$$\mathscr{B}_1 \times \mathscr{B}_2 \times \cdots \times \mathscr{B}_n.$$

Prove that

$$\mathscr{B}(\mathbb{R}^{d_1 + d_2 + \cdots + d_n}) = \mathscr{B}(\mathbb{R}^{d_1}) \times \mathscr{B}(\mathbb{R}^{d_2}) \times \cdots \times \mathscr{B}(\mathbb{R}^{d_n}).$$

[*Hint*: It suffices to discuss the case where $n = 2$. The general case will follow by induction. The class of all $E_1 \subset \mathbb{R}^{d_1}$ such that $E_1 \times \mathbb{R}^{d_2} \in \mathscr{B}(\mathbb{R}^{d_1+d_2})$ is a σ-algebra on $\mathbb{R}^{d_1}$ containing all open subsets of $\mathbb{R}^{d_1}$ and therefore contains $\mathscr{B}(\mathbb{R}^{d_1})$. This implies that

$$E_1 \in \mathscr{B}(\mathbb{R}^{d_1}) \to E_1 \times \mathbb{R}^{d_2} \in \mathscr{B}(\mathbb{R}^{d_1+d_2}).$$

Similarly we see that

$$E_2 \in \mathscr{B}(\mathbb{R}^{d_2}) \Rightarrow \mathbb{R}^{d_1} \times E_2 \in \mathscr{B}(\mathbb{R}^{d_1+d_2}).$$

Thus we have

$$E_i \in \mathscr{B}(\mathbb{R}^{d_i}) \qquad (i = 1, 2)$$
$$\Rightarrow E_1 \times E_2 = (E_1 \times \mathbb{R}^{d_2}) \cap (\mathbb{R}^{d_1} \times E_2) \in \mathscr{B}(\mathbb{R}^{d_1+d_2}),$$

which proves

$$\mathscr{B}(\mathbb{R}^{d_1}) \times \mathscr{B}(\mathbb{R}^{d_2}) \subset \mathscr{B}(\mathbb{R}^{d_1+d_2}).$$

The converse inclusion relation follows from the fact that every open subset of $\mathbb{R}^d$ is a countable union of sets of the form

$$(a_1, b_1) \times (a_2, b_2) \times \cdots \times (a_d, b_d).]$$

(ix) Let f be a map from S_1 into S_2 and $\mathscr{B}_i$ be a σ-algebra on S_i for $i = 1, 2$. f is called *measurable* $\mathscr{B}_1/\mathscr{B}_2$, written $f \in \mathscr{B}_1/\mathscr{B}_2$, if

$$E \in \mathscr{B}_2 \Rightarrow f^{-1}(E) \in \mathscr{B}_1.$$

Prove the following properties of measurability.

(a) If $f : S_1 \to S_2$ is measurable $\mathscr{B}_1/\mathscr{B}_2$ and $g : S_2 \to S_3$ is measurable $\mathscr{B}_2/\mathscr{B}_3$, then $g \circ f : S_1 \to S_3$ is measurable $\mathscr{B}_1/\mathscr{B}_3$.

(b) Let $\mathscr{B}_i$ be a σ-algebra on S_i for $i = 1, 2, \ldots, n$ and let π_i denote the i-projection from $S = S_1 \times S_2 \times \cdots \times S_n$ to S_i. Then

$$\pi_i \in \mathscr{B}_1 \times \mathscr{B}_2 \times \cdots \times \mathscr{B}_n | \mathscr{B}_i, \qquad i = 1, 2, \ldots, n.$$

If $\mathscr{B}$ is a σ-algebra on S such that $\pi_i \in \mathscr{B}|\mathscr{B}_i$ ($i = 1, 2, \ldots, n$), then $\mathscr{B} \supset \mathscr{B}_1 \times \mathscr{B}_2 \times \cdots \times \mathscr{B}_n$. Hence $\mathscr{B}_1 \times \mathscr{B}_2 \times \cdots \times \mathscr{B}_n$ is the least σ-algebra on S such that $\pi_i \in \mathscr{B}|\mathscr{B}_i$ ($i = 1, 2, \ldots, n$).

(x) Let S be a map from S_1 into S_2, $\mathscr{B}_1$ a σ-algebra on S_1, and $\mathscr{A}_2$ a class of subsets of S_2. If $f^{-1}(E) \in \mathscr{B}_1$ for every $E \in \mathscr{A}_2$, then f is measurable $\mathscr{B}_1/\sigma[\mathscr{A}_2]$.

(xi) Let $\mathscr{B}_1$ be a σ-algebra on S_1, S_2 a topological space, and f a map from S_1 into S_2. f is said to be measurable $\mathscr{B}_1$, written $f \in \mathscr{B}_1$, if $f \in \mathscr{B}_1/\mathscr{B}(S_2)$. Use (x) to prove that $f \in \mathscr{B}_1$ if $f^{-1}(G) \in \mathscr{B}_1$ for every open subset G of S_2.

(xii) Let S_1 and S_2 be topological spaces. A map $f: S_1 \to S_2$ is called a *Borel-measurable* map or a *Borel map* is $f \in \mathscr{B}(S_1)/\mathscr{B}(S_2)$. Prove that every continuous map is Borel-measurable.

(xiii) Let P be a probability measure on S_1, S_2 a topological space, and f a map from S_1 into S_2. f is called p-measurable if $f \in \mathscr{D}(P)/\mathscr{B}(S)$. Prove that if $A \in \mathscr{D}(P)$, then the indicator $1_A : S_1 \to \mathbb{R}^1$ is P-measurable.

(xiv) Let $\mathscr{B}, \mathscr{B}_1, \mathscr{B}_2, \ldots, \mathscr{B}_n$ be σ-algebras on $S, S_1, \ldots, S_n$, respectively, and f_i a map from S into S_i for $i = 1, 2, \ldots, n$. The map

$$f: S \to S_1 \times S_2 \times \cdots \times S_n, \qquad x \mapsto (f_1(x), f_2(x), \ldots, f_n(x))$$

is called the *product map* of $f_1, f_2, \ldots, f_n$. Prove that

$$f_i \in \mathscr{B}|\mathscr{B}_i \ (i = 1, 2, \ldots, n) \Rightarrow f \in \mathscr{B}|\mathscr{B}_1 \times \mathscr{B}_2 \times \cdots \times \mathscr{B}_n.$$

[*Hint*: Note that

$$\mathscr{B}' = \left\{ E \subset S_1 \times S_2 \times \cdots \times S_n | f^{-1}(E) \in \mathscr{B} \right\}$$

is a σ-algebra containing $E_1 \times E_2 \times \cdots \times E_n$ $(E_i \in \mathscr{B}_i, \ i = 1, 2, \ldots, n)$.]

(xv) A P-measurable set N with $P(N) = 0$ is called a *P-null set*. P is called a *complete probability measure* if every subset of every P-null set is a P-measurable (and so P-null) set. A probability space (Ω, P) is called a *complete probability space* if P is complete. Let P_1 and P_2 be two probability measures. If $\mathscr{D}(P_1) \subset \mathscr{D}(P_2)$ and $P_1(A) = P_2(A)$ for every $A \in \mathscr{D}(P_1)$, then P_2 is called an *extension* of P_1. If P_2 is complete, then P_2 is called a *complete extension*. Prove that every probability measure has the least complete extension, called the *Lebesgue extension*. [*Hint*: Define Q by

$$\mathscr{D}(Q) = \text{the class of all } A \text{ such that } B_1 \subset A \subset B_2$$

$$\text{and } P(B_2 - B_1) = 0 \text{ for some } B_1, B_2 \in \mathscr{D}(P);$$

$$Q(A) = P(B_1) \quad \text{for the above-mentioned } B_1.$$

Then Q is the Lebesgue extension of P.]

2.2 The extension theorem of probability measures

As we saw in the trial of throwing a die infinitely many times, we can easily define the probabilities of rather simple subsets of the sample space and hence obtain a set function defined on these sets. Now our problem is to extend this set function to a probability measure. We will prove a theorem that is useful for such problems. First we introduce some preliminary concepts.

Let $\mathscr{A}$ be a class of subsets of Ω. If $\mathscr{A}$ is closed under intersections, then $\mathscr{A}$ is called a *multiplicative class* on Ω. If Ω is closed under countable disjoint unions and proper differences and contains Ω, then $\mathscr{A}$ is called a *Dynkin class* on Ω. As we defined the σ-algebra generated by $\mathscr{A}$ (i.e., $\sigma[\mathscr{A}]$), we define the Dynkin class generated by $\mathscr{A}$, written $\delta[\mathscr{A}]$ to be the least Dynkin class that includes $\mathscr{A}$. Since every σ-algebra is a Dynkin class, it is obvious that

$$\delta[\mathscr{A}] \subset \sigma[\mathscr{A}].$$

Suppose that $\mathscr{A}$ is a Dynkin class and also is a multiplicative class. Since $\mathscr{A}$ contains Ω. We have

$$A \in \mathscr{A} \Rightarrow A^c = \Omega - A \in \mathscr{A},$$

and so

$$A_n \in \mathscr{A} \quad (n = 1, 2, \ldots)$$
$$\Rightarrow \bigcup_n A_n = \sum_n A_n \cap A_1^c \cap A_2^c \cap \ldots \cap A_{n-1}^c \in \mathscr{A}.$$

Therefore $\mathscr{A}$ is a σ-algebra.

Theorem 2.2.1 (The Dynkin class theorem). If $\mathscr{A}$ is multiplicative, then

$$\delta[\mathscr{A}] = \sigma[\mathscr{A}].$$

Proof: By the observation above it suffices to prove that $\delta[\mathscr{A}] \supset \sigma[\mathscr{A}]$. To do this it suffices to prove that $\delta[\mathscr{A}]$ is a σ-algebra. For this purpose it suffices to check that $\delta[\mathscr{A}]$ is a multiplicative class. Set

$$\mathscr{D}_1 = \{ A \mid A \cap B \in \delta[\mathscr{A}] \quad \text{for every } B \in \mathscr{A} \}.$$

Noting that $\delta[\mathscr{A}]$ is a Dynkin class, we can easily check that $\mathscr{D}_1$ is also a Dynkin class. Since $\mathscr{A}$ is multiplicative, $\mathscr{D}_1$ includes $\mathscr{A}$. Therefore $\mathscr{D}_1$ includes $\delta[\mathscr{A}]$ by the definition of $\delta[\mathscr{A}]$. This implies that

$$A \in \delta[\mathscr{A}], B \in \mathscr{A} \Rightarrow A \cap B \in \delta[\mathscr{A}].$$

Hence the class

$$\mathscr{D}_2 = \{ B \mid A \cap B \in \delta[\mathscr{A}] \quad \text{for every } A \in \delta[\mathscr{A}] \}$$

includes $\mathscr{A}$. As $\mathscr{D}_1$ is a Dynkin class, so is $\mathscr{D}_2$. Therefore we have $\delta[\mathscr{A}] \subset \mathscr{D}_2$. This implies that

$$A \in \delta[\mathscr{A}], \quad B \in \delta[\mathscr{A}] \Rightarrow A \cap B \in \delta[\mathscr{A}]. \qquad \blacksquare$$

Applying the Dynkin class theorem, we obtain the following:

Theorem 2.2.2 (The coincidence theorem of probability measures). Let P_1 and P_2 be probability measures on Ω and let $\mathscr{A}$ be a multiplicative class included in $\mathscr{D}(P_1) \cap \mathscr{D}(P_2)$. If P_1 and P_2 coincide with each other on $\mathscr{A}$, then they coincide on $\sigma[\mathscr{A}]$.

Proof: Consider the following class of sets:

$$\mathscr{B} = \{ A \in \mathscr{D}(P_1) \cap \mathscr{D}(P_2) \mid P_1(A) = P_2(A) \}.$$

Then $\mathscr{B}$ is a Dynkin class, since P is a probability measure. $\mathscr{B}$ includes $\mathscr{A}$ by the assumption. Hence $\mathscr{B}$ includes $\delta[\mathscr{A}]$. By the Dynkin class theorem, $\delta(\mathscr{A}) = \sigma[\mathscr{A}]$, since $\mathscr{A}$ is multiplicative. Thus we have $\mathscr{B} \supset \sigma[\mathscr{A}]$, which implies that $P_1(A) = P_2(A)$ for every $A \in \sigma[\mathscr{A}]$. $\blacksquare$

A probability measure on a topological space whose domain of definition is the Borel system on the space is called a *Borel probability measure*, whose Lebesgue extension is called a *regular probability measure*. Since the class of all open subsets of a topological space generates the Borel system, the coincidence theorem mentioned above shows that two Borel probability measures coincide with each other if they coincide on the class of all open subsets. Hence regular probability measures have the same property. In other words, Borell probability measures and regular probability measures are completely determined by their behaviors on the class of all open subsets. This fact holds true even if open subsets are replaced by closed subsets.

Let Ω be an arbitrary set. A set function μ^* defined for all subsets of Ω is called an *outer measure* on Ω, if it satisfies the following four conditions:

(C.1) $0 \leq \mu^*(A) \leq \infty$,

(C.2) $\mu^*(\phi) = 0$,

(C.3) $A \subset B \Rightarrow \mu^*(A) \leq \mu^*(B)$ (monotonicity),

(C.4) $\mu^*(\bigcup_{n=1}^{\infty} A_n) \leq \sum_{n=1}^{\infty} \mu^*(A_n)$ (subadditivity).

Let μ^* be an outer measure on Ω. A is called μ^*-*measurable*, if

$$\mu^*(W) = \mu^*(W \cap A) + \mu^*(W \cap A^c) \qquad \text{for every } W \subset \Omega.$$

Setting $A_{m+1} = A_{m+2} = \cdots = \varnothing$ in (C.4) and using (C.2), we obtain

$$\mu^*(A_1 \cup A_2 \cup \cdots \cup A_m) \leq \mu^*(A_1) + \mu^*(A_2) + \cdots + \mu^*(A_m).$$

Hence we have

$$\mu^*(W) \leqq \mu^*(W \cap A) + \mu^*(W \cap A^c),$$

which implies that the μ^*-measurability of A is characterized by the inequality

$$\mu^*(W) \geqq \mu^*(W \cap A) + \mu^*(W \cap A^c) \qquad (W \subset \Omega).$$

Theorem 2.2.3 (Carathéodory's theorem). Let μ^* be an outer measure of Ω and let $\mathscr{M}$ denote the class of all μ^*-measurable subsets. Then the restriction of μ^*, $\mu^*|_{\mathscr{M}}$, is a measure on Ω. If $\mu^*(\Omega) = 1$, then $\mu^*|_{\mathscr{M}}$ is a probability measure.

Proof: Denote $\mu^*|_{\mathscr{M}}$ by μ. Then $\mathscr{D}(\mu) = \mathscr{M}$. First we will prove that $\mathscr{M}$ is a σ-algebra. It is obvious that $\mathscr{M}$ contains Ω and is closed under complements. If $A, B \in \mathscr{M}$, then we have

$$\mu^*(W) \geqq \mu^*(W \cap A) + \mu^*(W \cap A^c)$$
$$\geqq \mu^*(W \cap A) + \mu^*(W \cap A^c \cap B) + \mu^*(W \cap A^c \cap B^c)$$
$$\geqq \mu^*(W \cap (A \cup B)) + \mu^*(W \cap (A \cup B)^c)$$
$$(\because A \cup B = A + A^c \cap B),$$

which implies that $A \cup B \in \mathscr{M}$. Hence $\mathscr{M}$ is closed under finite unions. This implies that $\mathscr{M}$ is closed under finite intersections, because $\mathscr{M}$ is closed under complements. Similarly, if $A_1, A_2, \ldots \in \mathscr{M}$ are disjoint, then

$$\mu^*(W) \geqq \mu^*(W \cap A_1) + \mu^*(W \cap A_1^c)$$
$$\geqq \mu^*(W \cap A_1) + \mu^*(W \cap A_1^c \cap A_2) + \mu^*(W \cap A_1^c \cap A_2^c)$$
$$= \mu^*(W \cap A_1) + \mu^*(W \cap A_2) + \mu^*(W \cap A_1^c \cap A_2^c)$$
$$\geqq \mu^*(W \cap A_1) + \mu^*(W \cap A_2) + \mu^*(W \cap A_1^c \cap A_2^c \cap A_3)$$
$$+ \mu^*(W \cap A_1^c \cap A_2^c \cap A_3^c)$$
$$= \mu^*(W \cap A_1) + \mu^*(W \cap A_2) + \mu^*(W \cap A_3)$$
$$+ \mu^*(W \cap A_1^c \cap A_2^c \cap A_3^c).$$

By induction we obtain

$$\mu^*(W) \geqq \sum_{n=1}^{N} \mu^*(W \cap A_n) + \mu^*(W \cap A_1^c \cap A_2^c \cap \cdots \cap A_N^c)$$
$$\geqq \sum_{n=1}^{N} \mu^*(W \cap A_n) + \mu^*\left(W \cap \bigcap_{n=1}^{\infty} A_n^c\right)$$

(by the monotonicity of μ).

Letting $N \to \infty$ we have

$$\mu^*(W) \geq \sum_{n=1}^{\infty} \mu^*(W \cap A_n) + \mu^*\left(W \cap \left(\sum_{n=1}^{\infty} A_n\right)^c\right)$$

$$\geq \mu^*\left(W \cap \sum_{n=1}^{\infty} A_n\right) + \mu^*\left(W \cap \left(\sum_{n=1}^{\infty} A_n\right)^c\right),$$

which implies that $\sum_{n=1}^{\infty} A_n \in \mathcal{M}$. Hence $\mathcal{M}$ is closed under disjoint countable unions. To deduce from this that $\mathcal{M}$ is closed under countable unions, note that

$$\bigcup_{n} A_n = \sum_{n=1}^{\infty} A_n \cap A_1^c \cap A_2^c \cap \cdots \cap A_{n-1}^c.$$

Thus we have proved that $\mathcal{M}$ is a σ-algebra. To complete the proof of our theorem it suffices to check the σ-additivity of μ. Setting

$$W = \sum_{n=1}^{\infty} A_n \qquad (A_1, A_2, \ldots \in \mathcal{M})$$

in the last inequality above, we obtain

$$\mu\left(\sum_{n=1}^{\infty} A_n\right) \geq \sum_{n=1}^{\infty} \mu(A_n),$$

which, combined with the subadditivity of μ^*, implies

$$\mu\left(\sum_{n=1}^{\infty} A_n\right) = \sum_{n=1}^{\infty} \mu(A_n). \qquad \blacksquare$$

A class of subsets of a set Ω is called an *algebra* of subsets of Ω or an algebra on Ω if it contains Ω and is closed under finite unions and complements. Let $\mathcal{A}$ be an algebra on Ω. A set function p defined for all sets in $\mathcal{A}$ is called an *elementary probability measure* on Ω (or on $\mathcal{A}$), if it satisfies the following conditions:

(p.1) $p(A) \geq 0$,

(p.2) $p(A + B) = p(A) + p(B)$ (additivity),

(p.3) $p(\Omega) = 1$.

The algebra $\mathcal{A}$ is the *domain* (of definition) of this set function p. The probability measures on a finite set Ω observed in Chapter 1 are elementary probability measures on Ω with domain 2^Ω. It is easy to check that elementary probability measures enjoy all properties of P discussed in Section 1.1. Needless to say, every σ-algebra is an algebra and every probability measure is an elementary probability measure.

The class of all subsets of $[0, 1]$ represented as finite disjoint unions of subintervals of $[0, 1]$ is an algebra on $[0, 1]$. If we associate every element of this algebra with the sum of the lengths of the intervals constituting it, we obtain an elementary probability measure. Similarly the set function $P(I)$, $I \in \mathscr{I}$, defined on the sample space for the trial of tossing a coin infinitely many times (Section 2.1) can be extended to an elementary probability measure on the algebra that consists of all finite disjoint unions of elements of $\mathscr{I}$. The following theorem is useful to extend an elementary probability measure to a probability measure.

Theorem 2.2.4 (The extension theorem of probability measures). Let $\mathscr{A}$ be an algebra on Ω and let p be an elementary probability measure on $\mathscr{A}$. A necessary and sufficient condition for extendability of p to a probability measure on $\sigma[\mathscr{A}]$ is that p be σ-additive on $\mathscr{A}$, namely, that if $A_n \in A$ $(n = 1, 2, \ldots)$ are disjoint and $\sum_{n=1}^{\infty} A_n \in \mathscr{A}$, then $p(\sum_{n=1}^{\infty} A_n) = \sum_{n=1}^{\infty} p(A_n)$. This extension is unique under this condition.

Remark: This condition is equivalent to each one of the following conditions:
   (a) $A_1 \supset A_2 \supset \cdots, A_n \in \mathscr{A}(n = 1, 2, \ldots), \bigcap_{n=1}^{\infty} A_n = \emptyset$
      $\Rightarrow \lim_{n \to \infty} p(A_n) = 0$.
   (b) $A_1 \supset A_2 \supset \cdots, A_n \in \mathscr{A}(n = 1, 2, \ldots)$, $\inf p(A_n) > 0$
      $\Rightarrow \bigcap_{n=1}^{\infty} A_n \neq \emptyset$ (The intersection property of p).

Proof of Theorem 2.2.4: The uniqueness of the extension follows immediately from the coincidence theorem (Theorem 2.2.2). The condition is clearly necessary. We will construct a probability measure on $\sigma[\mathscr{A}]$ that coincides with p on $\mathscr{A}$ under this condition. Define an outer measure μ^* by

$$\mu^*(B) = \inf\left\{ \sum_{n=1}^{\infty} p(A_n) \,\middle|\, \bigcup_{n=1}^{\infty} A_n \supset B, A_n \in \mathscr{A}(n = 1, 2, \ldots) \right\}.$$

It is easy to see that the set function μ^* thus defined is an outer measure with $\mu^*(\Omega) = 1$. The theorem of Carathéodory ensures that the class of all μ^*-measurable sets, $\mathscr{M}$, is a σ-algebra and that $\mu = \mu^*|_{\mathscr{M}}$ is a probability measure. Therefore if we can prove that

$$\mathscr{A} \subset \mathscr{M}, \qquad \mu^*(A) = p(A) \qquad (A \in \mathscr{A}),$$

then the restriction $P = \mu|_{\sigma[A]}$ is the extension to be sought. Note that $\mathscr{A} \subset \mathscr{M}$ implies that $\mathscr{D}(\mu) = \mathscr{M} \supset \sigma[\mathscr{A}]$.

We will first observe that p is *σ-subadditive* on $\mathscr{A}$, namely,

$$A \in \mathscr{A}, \qquad A_n \in \mathscr{A} \quad (n = 1, 2, \ldots), \qquad A \subset \bigcup_{n=1}^{\infty} A_n$$

$$\Rightarrow A = \sum_{n=1}^{\infty} A \cap \left(A_n \setminus \bigcup_{k=1}^{n-1} A_k \right)$$

$$\Rightarrow p(A) = \sum_{n=1}^{\infty} p\left(A \cap \left(A_n \setminus \bigcup_{k=1}^{n-1} A_k \right) \right)$$

$$\Rightarrow p(A) \leq \sum_{n=1}^{\infty} p(A_n),$$

which implies $p(A) \leq \mu^*(A)$. Since

$$A = A \cup \varnothing \cup \varnothing \cup \cdots$$

implies

$$\mu^*(A) \leq p(A) + p(\varnothing) + p(\varnothing) + \cdots = p(A),$$

we obtain $\mu^*(A) = p(A)$ for every $A \in \mathscr{A}$.

It remains only to prove that $\mathscr{A} \subset \mathscr{M}$. Let A be any set in the class $\mathscr{A}$ and W any subset of Ω. Consider an arbitrary covering of W by $A_n \in \mathscr{A}$, $n = 1, 2, \ldots$. Then

$$W \cap A \subset \bigcup_{n=1}^{\infty} A_n \cap A \qquad \text{and} \qquad W \cap A^c \subset \bigcup_{n=1}^{\infty} A_n \cap A^c.$$

Hence we obtain

$$\mu^*(W \cap A) \leq \sum_{n=1}^{\infty} p(A_n \cap A),$$

and $\quad \mu^*(W \cap A^c) \leq \sum_{n=1}^{\infty} p(A_n \cap A^c),$

Therefore

$$\mu^*(W \cap A) + \mu^*(W \cap A^c) \leq \sum_{n=1}^{\infty} p(A_n).$$

Taking the infimum of the right-hand side, we obtain

$$\mu^*(W \cap A) + \mu^*(W \cap A^c) \leq \mu^*(W).$$

namely, $A \in \mathscr{M}$. Thus $\mathscr{A} \subset \mathscr{M}$ is proved. ∎

As an application of this theorem we will define *Lebesgue–Stieltjes measures*. Let P be a Borel probability measure on $\mathbb{R}^1$. The function

$$F(x) = F_P(x) = P((-\infty, x]), \qquad x \in \mathbb{R}^1$$

is called the *distribution function* of P. It is obvious that $F(x)$ satisfies the following three conditions:

(F.1) $x \leq y \Rightarrow F(x) \leq F(y)$ (monotonicity),

(F.2) $F(x+) = F(x)$ $(F(x+) = \lim_{y \downarrow x} F(y))$ (right-continuity),

(F.3) $F(-\infty) = 0$, $F(\infty) = 1$,

  $(F(-\infty) = \lim_{x \downarrow -\infty} F(x)$, $F(\infty) = \lim_{x \uparrow \infty} F(x))$.

Conversely, if $F(x)$ satisfies these conditions, then there exists a unique Borel probability measure P whose distribution function is $F(x)$, which is called the Lebesgue–Stieltjes measure corresponding to $F(x)$. The existence can be proved by using the extension theorem as follows. Set

$$p((a, b]) = F(b) - F(a), \qquad -\infty \leq a < b \leq \infty$$

for every left-open interval $(a, b]$ where $(a, \infty]$ is replaced by (a, ∞). Let $\mathscr{A}$ be the class of all elementary sets, where an *elementary set* is defined to be a finite disjoint union of left-open intervals. $\mathscr{A}$ is an algebra on $\mathbb{R}^1$ and p can be extended to an elementary probability measure in the obvious way. If we can prove the intersection property of p, then p is extended to a Borel probability measure, whose Lebesgue extension is the regular probability measure to be sought. Suppose that

$$A_1 \supset A_2 \supset \cdots, \qquad A_n \in \mathscr{A} \quad (n = 1, 2, \ldots), \qquad \alpha = \inf p(A_n) > 0.$$

Since F is right-continuous, we can take, for any left-open interval $I = (a, b]$, a left-open interval J such that the closure of J, written $\bar{J}$, is included in I and that $p(I) - p(J)$ is arbitrarily small. In case I is an infinite interval, we can take a bounded left-open interval by virtue of condition (F.3). Using this fact, we can take a bounded elementary set B_n for each A_n such that

$$\bar{B}_n \subset A_n \qquad \text{and} \qquad p(A_n) - p(B_n) < 2^{-n-1}\alpha, \, n = 1, 2, \ldots.$$

Then we have

$$p(A_n) - p\left(\bigcap_{i=1}^{n} B_i \right) = p\left(\bigcup_{i=1}^{n} (A_n - B_i) \right) \leq p\left(\bigcup_{i=1}^{n} (A_i - B_i) \right)$$

$$\leq \sum_{i=1}^{n} p(A_i - B_i) < \sum_{i=1}^{n} 2^{-i-1}\alpha < \alpha/2.$$

Since $p(A_n) \geq \alpha$, we have

$$p\left(\bigcap_{i=1}^{n} B_i \right) > \alpha/2,$$

which implies that

$$\bigcap_{i=1}^{n} \bar{B}_i \supset \bigcap_{i=1}^{n} B_i \neq \emptyset, \qquad n = 1, 2, \ldots .$$

Observing that the $\bar{B}_i$ $(i = 1, 2, \ldots)$ are bounded closed sets, we can use Cantor's intersection theorem to obtain

$$\bigcap_{n=1}^{\infty} A_n \supset \bigcap_{n=1}^{\infty} \bar{B}_n \neq \emptyset,$$

which proves the intersection property of p. This completes the proof of the existence of P. The uniqueness is obvious by the coincidence theorem. Note that the left-open intervals form a multiplicative class generating the Borel system on $\mathbb{R}^1$.

Exercise 2.2. Prove that the set function $P(I)$, $I \in \mathscr{I}$, observed for the trial of tossing a coin infinitely many times (Section 2.1) can be extended uniquely to a probability measure on $\sigma(\mathscr{I})$.

[*Hint*: We can extend $P(I)$ to an elementary probability measure on the algebra $\mathscr{A}$ that consists of all finite disjoint unions of elements of $\mathscr{I}$. To verify the intersection property of this elementary measure, observe that

$$\varphi \colon \Omega \to \mathbf{K} \quad \text{(Cantor set)}, \qquad (\omega_1, \omega_2, \ldots) \mapsto \sum_{n=1}^{\infty} \frac{2\omega_n}{3^n}$$

is a bijective map and that $\varphi(A)$ is a closed subset of K for every $A \in \mathscr{A}$.]

2.3 Direct products of probability measures

Direct products of probability measures are used to make a mathematical formulation of direct composite trials and independent random variables, as we have seen in Chapter 1 for finite trials.

Let us start with the definition of the direct product of two probability measures. Let P_1 and P_2 be probability measures on Ω_1 and Ω_2, respectively, and denote $\Omega_1 \times \Omega_2$ by Ω. A probability measure P on Ω with $\mathscr{D}(P) = \mathscr{D}(P_1) \times \mathscr{D}(P_2)$ is called the direct product of P_1 and P_2 (written $P_1 \times P_2$) if it satisfies

$$P(B_1 \times B_2) = P_1(B_1) P_2(B_2), \qquad B_i \in \mathscr{D}(P_i); i = 1, 2.$$

The probability space (Ω, P) is called the direct product of (Ω_1, P_1) and (Ω_2, P_2), written

$$(\Omega, P) = (\Omega_1, P_1) \times (\Omega_2, P_2).$$

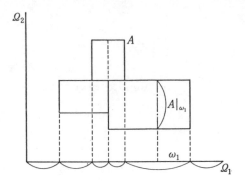

Figure 2.1

For example, the Lebesgue measure on $[0, 1]^2$ is the direct product of that on $[0, 1]$ and itself. In Chapter 1 we defined $P = P_1 \times P_2$ by $P\{(\omega_1, \omega_2)\} = P_1\{\omega_1\} P_2\{\omega_2\}$ in case Ω_1 and Ω_2 are finite sets. But the above-mentioned example shows that this definition is not adequate for general probability measures.

We will prove the existence and the uniqueness of $P = P_1 \times P_2$ for P_1 and P_2 given. The class $\mathscr{I}$ of all subsets of Ω of the form

$$B_1 \times B_2, \qquad B_i \in \mathscr{D}(P_i); \, i = 1, 2,$$

is a multiplicative class generating $\mathscr{B} = \mathscr{D}(P_1) \times \mathscr{D}(P_2)$. Hence the uniqueness is obvious by the coincidence theorem of probability measures. To prove the existence we will use the extension theorem of probability measures. First we define $p: \mathscr{I} \to [0, 1]$ by

$$p(B_1 \times B_2) = P_1(B_1) P_2(B_2)$$

and extend this to an elementary probability measure on the algebra

$$\mathscr{A} = \left\{ A = \sum_{i=1}^{n} I_i \middle| I_i \in \mathscr{I}, n = 1, 2, \dots \right\}.$$

If we verify the intersection property of p, then p can be extended to a probability measure, which turns out to be the product measure $P_1 \times P_2$.

Every set A in the algebra $\mathscr{A}$ can be written in the form:

$$A = \sum_{i=1}^{n} B_{1,i} \times B_{2,i}, \qquad B_{1,i} \in \mathscr{D}(P_1); \, B_{2,i} \in \mathscr{D}(P_2); \, i = 1, 2, \dots, n,$$

by the definition. By taking subdivisions and unions we can assume that $B_{1,i}$, $i = 1, 2, \dots, n$ are disjoint and $\Omega_1 = \sum_i B_{1,i}$, where $B_{2,i}$ may be an empty set for some i. (See Figure 2.1.) Then it is obvious that

$$\{\omega_2 \in \Omega_2 | (\omega_1, \omega_2) \in A\} = B_{2,i} \qquad \text{for } \omega_1 \in B_{1,i}.$$

(The left-hand side is called the *section* of A by $\omega_1 \in \Omega_1$, written $A|_{\omega_1}$.) If $p(A) \geq \alpha > 0$, then the P_1-measure of the following set $A_1(\alpha) \subset \Omega_1$ is no less than $\alpha/2$:

$$A_1(\alpha) = \sum B_{1,i}, \qquad P_2(B_{2,i}) \geq \alpha/2,$$

where $\sum$ ranges over all indexes i with $P_2(B_{2,i}) \geq \alpha/2$. To prove this we change the numbering so that $\sum$ ranges over $i = 1, 2, \ldots, r$. Then we have

$$\alpha \leq p(A) = \sum_{i=1}^{r} P_1(B_{1,i}) P_2(B_{2,i}) + \sum_{i=r+1}^{n} P_1(B_{1,i}) P_2(B_{2,i})$$

$$\leq \sum_{i=1}^{r} P_1(B_{1,i}) \cdot 1 + \sum_{i=r+1}^{n} P_1(B_{1,i}) \alpha/2$$

$$\leq P_1(A_1(\alpha)) + P_1(A_1(\alpha)^c) \alpha/2$$

$$\leq P_1(A_1(\alpha)) + \alpha/2$$

and so

$$P_1(A_1(\alpha)) \geq \alpha/2.$$

Keeping this in mind, we will verify the intersection property of p. Suppose that

$$A' \supset A'' \supset \cdots \supset A^{(n)} \supset \cdots \qquad \text{and} \qquad p(A^{(n)}) \geq \alpha > 0.$$

Then it is obvious that

$$A_1'(\alpha) \supset A_1''(\alpha) \supset A_1'''(\alpha) \supset \cdots$$

By the observation above we can see that all these sets have P_1-measure $\geq \alpha/2$. Hence

$$P_1\left(\bigcap_{n=1}^{\infty} A_1^{(n)}(\alpha) \right) = \lim_{n \to \infty} P_1(A_1^{(n)}(\alpha)) \geq \alpha/2,$$

so this intersection contains at least one point, say ω_1. Then the section

$$A^{(n)}|_{\omega_1} = \{ \omega_2 \in \Omega_2 | (\omega_1, \omega_2) \in A^{(n)} \}$$

has P_2-measure $\geq \alpha/2$ for every n. As $\{A^{(n)}\}$ is decreasing, so is $\{A^{(n)}|_{\omega_1}\}$. In the same way as above we can find a common point ω_2 of these sections. Then $(\omega_1, \omega_2) \in \bigcap_n A^{(n)}$. This completes the proof of the existence of $P_1 \times P_2$.

The direct product of more than two probability measures is defined by

$$P_1 \times P_2 \times P_3 = (P_1 \times P_2) \times P_3,$$

$$P_1 \times P_2 \times P_3 \times P_4 = (P_1 \times P_2 \times P_3) \times P_4,$$

$$\vdots$$

It is obvious that $P_1 \times P_2 \times \cdots \times P_n$ is characterized by

$$\mathscr{D}(P_1 \times P_2 \times \cdots \times P_n) = \mathscr{D}(P_1) \times \mathscr{D}(P_2) \times \cdots \times \mathscr{D}(P_n),$$

$$(P_1 \times P_2 \times \cdots \times P_n)(B_1 \times B_2 \times \cdots \times B_n) = P_1(B_1)P_2(B_2) \cdots P_n(B_n),$$

$$B_i \in \mathscr{D}(P_i), \qquad i = 1, 2, \ldots, n.$$

We often denote $P_1 \times P_2 \times \cdots \times P_n$ by $\Pi_{i=1}^n P_i$. The direct product of probability spaces is defined by

$$\prod_{i=1}^n (\Omega_i, P_i) = \left(\prod_{i=1}^n \Omega_i, \prod_{i=1}^n P_i \right) \qquad \left(\prod_{i=1}^n \Omega_i = \Omega_1 \times \Omega_2 \times \cdots \times \Omega_n \right).$$

Let us consider the direct product of a sequence of probability measures. Let P_n be a probability measure on Ω_n for $n = 1, 2, 3, \ldots$ and let Ω denote the Cartesian product of Ω_n ($n = 1, 2, \ldots$); that is,

$$\Omega = \prod_{n=1}^{\infty} \Omega_n = \Omega_1 \times \Omega_2 \times \cdots \times \Omega_n \times \cdots$$

Let $\pi_n \colon \Omega \to \Omega_n$ be the n-projection. The σ-algebra on Ω generated by

$$\pi_n^{-1}(B_n), \qquad B_n \in \mathscr{D}(P_n), n = 1, 2, \ldots$$

is called the product σ-algebra of $\mathscr{D}(P_n)$ ($n = 1, 2, \ldots$), written

$$\prod_{n=1}^{\infty} \mathscr{D}(P_n) \qquad \text{or} \qquad \mathscr{D}(P_1) \times \mathscr{D}(P_2) \times \cdots.$$

This definition is available to the product σ-algebra of an arbitrary family of σ-algebras. A probability measure P on Ω is called the direct product of P_n ($n = 1, 2, \ldots$), written

$$P = \prod_{n=1}^{\infty} P_n \qquad \text{or} \qquad P_1 \times P_2 \times \cdots$$

if P has the following properties:

$$\mathscr{D}(P) = \prod_{n=1}^{\infty} \mathscr{D}(P_n),$$

$$P\left(\bigcap_{i=1}^n \pi_i^{-1}(B_i) \right) = \prod_{i=1}^n P_i(B_i),$$

$$B_i \in \mathscr{D}(P_i); i = 1, 2, \ldots, n; n = 1, 2, \ldots.$$

Our task is to prove the existence and uniqueness of P. Let $\mathscr{I}$ denote the class of all subsets of Ω of the form:

$$\bigcap_{i=1}^n \pi_i^{-1}(B_i) \qquad \text{or} \qquad B_1 \times B_2 \times \cdots \times B_n \times \Omega_{n+1} \times \Omega_{n+2} \times \cdots.$$

Since $\mathscr{I}$ is a multiplicative class generating the product σ-algebra $\prod_{n=1}^{\infty} \mathscr{D}(P_n)$, the uniqueness is obvious by the coincidence theorem of probability measures.

Now we will prove the existence of P, using the extension theorem. Denote by $\mathscr{A}$ the class of all finite disjoint unions of sets in $\mathscr{I}$. Then $\mathscr{A}$ is an algebra on Ω generating $\mathscr{D}(P)$. We define $P(B)$ for $B \in \mathscr{I}$ by the second property of P mentioned above; that is,

$$P(B_1 \times B_2 \times \cdots \times B_n \times \Omega_{n+1} \times \Omega_{n+2} \times \cdots)$$
$$= P_1(B_1)P_2(B_2)\ldots P_n(B_n).$$

Note that this holds true even if $B_i = \Omega_i$ for some i, because $P(\Omega_i) = 1$. The set function P on $\mathscr{I}$ can be extended to an elementary probability measure on $\mathscr{A}$. It remains only to verify the intersection property of this elementary probability measure.

Let $\{A^{(n)}, \ n = 1, 2, \ldots\} \subset \mathscr{A}$ be a decreasing sequence satisfying $P(A^{(n)}) > \alpha > 0 \ (n = 1, 2, \ldots)$. We will find a point $\omega = (\omega_1, \omega_2, \ldots) \in \Omega$ which belongs to $A^{(n)}$ for every n. In the same way as we defined A and P on $\Omega = \Omega_1 \times \Omega_2 \times \cdots$ we define $\mathscr{A}'$ and P' on $\Omega' = \Omega_2 \times \Omega_3 \times \cdots$. Then every $A \in \mathscr{A}$ is represented by

$$A = \sum_{i=1}^{n} B_{1,i} \times B_{2,i}, \qquad B_{1,i} \in \mathscr{D}(P_1); \ B_{2,i} \in \mathscr{A}'; \ \Omega_1 = \sum_{i=1}^{n} B_{1,i},$$

and it holds that

$$P(A) = \sum_{i=1}^{n} P_1(B_{1,i}) P'(B_{2,i}).$$

Making use of the above-mentioned argument in the product of two probability measures, we can find $\omega_1 \in \Omega_1$ such that

$$P'\left(A^{(n)}|_{\omega_1}\right) \geq \alpha/2, \qquad n = 1, 2, \ldots,$$

so

$$A^{(n)}|_{\omega_1} \neq \varnothing, \qquad n = 1, 2, \ldots,$$

where $A^{(n)}|_{\omega_1}$ is the section of $A^{(n)}$ by ω_1, namely

$$A^{(n)}|_{\omega_1} = \{(\omega_2, \omega_3, \ldots) | (\omega_1, \omega_2, \ldots) \in A^{(n)}\} \in \mathscr{A}'.$$

Since $\{A^{(n)}|_{\omega_1}, \ n = 1, 2, \ldots\}$ is a decreasing sequence and $P(A^{(n)}|_{\omega_1}) \geq \alpha/2$, we can repeat the same argument to find $\omega_2 \in \Omega_2$ such that

$$P''\left(A^{(n)}|_{(\omega_1, \omega_2)}\right) \geq \alpha/4, \qquad n = 1, 2, \ldots,$$

so

$$A^{(n)}|_{(\omega_1, \omega_2)} \neq \varnothing, \qquad n = 1, 2, \ldots .$$

Repeating this, we can find $\omega_k \in \Omega_k$, $k = 1, 2, \ldots$ such that

$$A^{(n)}|_{(\omega_1, \omega_2, \ldots, \omega_k)} \neq \varnothing, \qquad n = 1, 2, \ldots, \quad k = 1, 2, \ldots$$

We will prove that

$$\omega \equiv (\omega_1, \omega_2, \ldots) \in A^{(n)}, \qquad n = 1, 2, \ldots$$

Since $A^{(n)}$ belongs to $\mathscr{A}$, it is represented in the form:

$$A^{(n)} = B \times \Omega_{N+1} \times \Omega_{N+2} \times \cdots, \qquad B \subset \Omega_1 \times \Omega_2 \times \cdots \times \Omega_N,$$

where N depends on n obviously. Since $A^{(n)}|_{(\omega_1, \omega_2, \ldots, \omega_N)} \neq \varnothing$, we have $(\omega_1, \omega_2, \ldots, \omega_N) \in B$, which implies that

$$\omega = (\omega_1, \omega_2, \ldots, \omega_N, \omega_{N+1}, \omega_{N+2}, \ldots) \in B \times \Omega_{N+1} \times \Omega_{N+2} \times \cdots = A^{(n)}.$$

This completes the proof of the existence of $\prod_{n=1}^{\infty} P_n$. ∎

We can also define the product of general measures. A measure m on Ω is called a finite measure if $m(\Omega) < \infty$, and a σ-finite measure if there exists a sequence $\{A_n\}$ such that

$$\Omega = \sum_{n=1}^{\infty} A_n, \qquad m(A_n) < \infty; n = 1, 2, \ldots$$

If m is a finite measure on Ω with $m(\Omega) > 0$, then $P(A) = m(A)/m(\Omega)$ is a probability measure. Therefore, we define

$$m_1 \times m_2 = m_1(\Omega_1) m_2(\Omega_2) \frac{m_1}{m_1(\Omega_1)} \times \frac{m_2}{m_2(\Omega_2)},$$

if both $m(\Omega_1)$ and $m(\Omega_2)$ do not vanish, and $m_1 \times m_2 \equiv 0$ if otherwise. Then we have

$$(m_1 \times m_2)(A_1 \times A_2) = m_1(A_1) m_2(A_2),$$

as we desire. Similarly, we define the product of a finite number of measures. If $\{m_n\}$ is a sequence of measures, we define

$$\prod_{n=1}^{\infty} m_n = \prod_{n=1}^{\infty} m_n(\Omega_n) \prod_{n=1}^{\infty} \frac{m_n}{m_n(\Omega_n)},$$

provided

$$\sum_{n=1}^{\infty} |m_n(\Omega_n) - 1| < \infty.$$

This condition is necessary and sufficient for $\prod_{n=1}^{\infty} m(\Omega_n)$ to be finite.

If m_1 and m_2 are σ-finite measures on Ω_1 and Ω_2, respectively, then we can decompose Ω_i as follows:

$$\Omega_i = \sum_{n=1}^{\infty} A_{in}, \qquad m_i(A_{in}) < \infty; n = 1, 2, \ldots; i = 1, 2.$$

Then

$$m_{in}(B_i) = m_i(B_i \cap A_{in}), \qquad i = 1, 2,$$

are finite measures on Ω_i, $i = 1$ and 2, respectively, and $m_{1n} \times m_{2k}$ are defined for $n, k = 1, 2, \ldots$. Hence we define

$$(m_1 \times m_2)(A) = \sum_{n,k=1}^{\infty} (m_{1n} \times m_{2k})(A).$$

Similarly we can define the product of a finite number of measures. But the product of a sequence of general measures cannot always be defined.

Let $\{T_n\}$ be a finite or infinite sequence of trials and let (Ω_n, P_n) be the probability space of T_n for $n = 1, 2, \ldots$. Similarly to the case of finite trials the probability space (Ω, P) for the direct composition of $\{T_n\}$ is given by

$$(\Omega, P) = \prod_n (\Omega_n, P_n).$$

Exercise 2.3.

(i) Suppose that we throw a die infinitely many times. Find the probability p_n that 6 turns up exactly n times before 5 appears. [*Hint*: Determine the probability space (Ω, P) for the trial and the subset A_n of Ω corresponding to the event in question and then compute $p_n = P(A_n)$. Note that the product of a sequence of probability measures is necessary for such a simple problem.]

(ii) Let $\overline{P}$ denote the Lebesgue extension of a probability measure P. Prove that

$$\overline{P_1 \times P_2 \times \cdots} = \overline{\overline{P}_1 \times \overline{P}_2 \times \cdots}.$$

($\overline{P_1 \times P_2 \times \cdots}$ is called the *complete direct product* of $\{P_n\}$.)

(iii) Let $\mathcal{B}_i$ be a σ-algebra on Ω_i for $i = 1, 2$. Prove that the section of $B \in \mathcal{B}_1 \times \mathcal{B}_2$ by $\omega_1 \varepsilon \Omega_1$ belongs to $\mathcal{B}_2$. [*Hint*: The class of all sets B with this property is a σ-algebra on $\Omega_1 \times \Omega_2$ including $B_1 \times B_2$ ($B_i \in \mathcal{B}_i$, $i = 1, 2$).]

2.4   Standard probability spaces

Let (Ω_i, P_i), $i = 1, 2$, be probability spaces. A bijective map f: $\Omega_1 \to \Omega_2$ (i.e., a 1-to-1 map from Ω_1 onto Ω_2) is called a *strictly isomorphic map* from (Ω_1, P_1) to (Ω_2, P_2), if

$$A_1 \in \mathscr{D}(P_1) \Leftrightarrow f(A_1) \in \mathscr{D}(P_2) \Rightarrow P_1(A_1) = P_2(f(A_1)).$$

If there exists such a map f, then (Ω_1, P_1) is said to be strictly isomorphic to (Ω_2, P_2), written

$$(\Omega_1, P_1) \approx (\Omega_2, P_2) \qquad \text{or} \qquad (\Omega_1, P_1) \approx (\Omega_2, P_2) \, (f).$$

Strict isomorphism is an equivalence relation, because we have

$$(\Omega, P) \approx (\Omega, P) \, (I) \, (I \text{: identity map}) \qquad \text{(reflexivity)};$$

$$(\Omega_1, P_1) \approx (\Omega_2, P_2) \, (f) \Rightarrow (\Omega_2, P_2) \approx (\Omega_1, P_1) \, (f^{-1}) \qquad \text{(symmetry)};$$

$$(\Omega_1, P_1) \approx (\Omega_2, P_2) \, (f), (\Omega_2, P_2) \approx (\Omega_3, P_3) \, (g)$$

$$\Rightarrow (\Omega_1, P_1) \approx (\Omega_3, P_3) \, (g \circ f) \qquad \text{(transitivity)}.$$

We introduce isomorphism as a relation weaker than strict isomorphism. Let Ω' be a subset of a probability (Ω, P) with $P(\Omega') = 1$ and P' denote the restriction of P to Ω', written $P|_{\Omega'}$, namely,

$$\mathscr{D}(P') = \{ A' \in \mathscr{D}(P) \, | \, A' \subset \Omega' \},$$

$$P'(A') = P(A'), \qquad A' \in \mathscr{D}(P').$$

Then P' is a probability measure on Ω'. The probability space (Ω', P') is called the *restriction* of (Ω, P) to Ω', written $(\Omega, P)|_{\Omega'}$. We define (Ω_1, P_1) to be *isomorphic* to (Ω_2, P_2), written

$$(\Omega_1, P_1) \sim (\Omega_2, P_2),$$

if we can find $\Omega_i' \subset \Omega_i$ with $P_i(\Omega_i') = 1$ for $i = 1, 2$ such that $(\Omega_1, P_1)|_{\Omega_1'}$ is strictly isomorphic to $(\Omega_2, P_2)|_{\Omega_2'}$. Isomorphism is also an equivalence relation. Reflexivity and symmetry are obvious. Suppose that

$$(\Omega_1, P_1) \sim (\Omega_2, P_2), \qquad (\Omega_2, P_2) \sim (\Omega_3, P_3).$$

Then we have

$$\left(\Omega_1', P_1' \right) \approx \left(\Omega_2', P_2' \right) (f), \qquad \left(\Omega_2'', P_2'' \right) \approx \left(\Omega_3', P_3' \right) (g),$$

where (Ω_1', P_1'), (Ω_2', P_2'), (Ω_2'', P_2''), and (Ω_3', P_3') are restrictions of (Ω_1, P_1), (Ω_2, P_2), (Ω_2, P_2), and (Ω_3, P_3), respectively. If we set

$$\tilde{\Omega}_2 = \Omega_2' \cap \Omega_2'', \qquad \tilde{\Omega}_1 = f^{-1}(\tilde{\Omega}_2), \qquad \tilde{\Omega}_3 = g(\tilde{\Omega}_2),$$

then $P_i(\tilde{\Omega}_i) = 1$ $(i = 1, 2, 3)$ and

$$(\tilde{\Omega}_1, \tilde{P}_1) \approx (\tilde{\Omega}_2, \tilde{P}_2) \approx (\tilde{\Omega}_3, \tilde{P}_3),$$

where $\tilde{P}_i = P_i|_{\tilde{\Omega}_i}$ $(i = 1, 2, 3)$. Hence it follows that $(\Omega_1, P_1) \sim (\Omega_3, P_3)$. ∎

Definition 2.4. If (Ω, P) is isomorphic to $(\mathbb{R}^1, \mu)$, μ being a regular probability measure on $\mathbb{R}^1$, then (Ω, P) is called a *standard probability space* and P is called a *standard probability measure* on Ω.

All probability spaces appearing in practical applications are standard, as we shall see later.

Theorem 2.4.1. Let S be a complete separable metric space. Then every regular probability measure P on S is standard.

Before beginning the proof of this theorem we will prove some preliminary facts. Let P be a regular probability measure on a topological space S. A P-measurable set A is called *K-regular* if for every $\varepsilon > 0$ we can find a compact set $K \subset A$ such that $P(A - K) < \varepsilon$. If every P-measurable subset of S is K-regular, then P is called K-regular.

Lemma 2.4.1. Let S be a complete separable metric space. Then every regular probability measure on S is K-regular.

Proof: Let ρ denote the metric on S. S being separable, we can find a sequence $\{a_n\}$ dense in S. Then we have

$$S = \bigcup_{n=1}^{\infty} B_{nk}, \qquad B_{nk} = \{x \in S | \rho(x, a_n) \leq 1/k\}$$

and so

$$\bigcup_{n=1}^{N} B_{nk} \uparrow S \qquad (N \to \infty).$$

Hence we can take a sufficiently large $N(k)$ such that

$$P(S - B_k) < 2^{-(k+1)}\varepsilon, \qquad B_k = \bigcup_{n=1}^{N(k)} B_{nk}.$$

Since B_k is closed,

$$K = \bigcap_{k=1}^{\infty} B_k$$

is also closed. Since

$$K \subset B_k = \bigcup_{n=1}^{N(k)} B_{nk} \quad \text{for every } k$$

and the diameter of B_{nk} is no more than $2/k$, K is totally bounded. Hence K is compact, being a totally bounded closed subset of a complete metric space S. Observing that

$$P(S-K) = P\left(S - \bigcap_k B_k\right) = P\left(\bigcup_k (S-B_k)\right)$$

$$\leq \sum_k P(S-B_k) < \varepsilon,$$

we can see that S is K-regular.

Let A be a Borel subset of S. Then for every $\varepsilon > 0$ we can find a closed set $F \subset A$ and an open set $G \supset A$ such that $P(G-F) < \varepsilon$. To prove this it suffices to show that the class $\mathscr{B}$ of all A satisfying this property includes $\mathscr{B}(S)$. Since every open subset of a metric space is represented as a countable union of closed sets, $\mathscr{B}$ contains all open sets. $\mathscr{B}$ is closed under complements, because $G \supset A \supset F$ implies $F^c \supset A^c \supset G^c$. For $A_n \in \mathscr{B}$ $(n = 1, 2, \ldots)$ we take F_n (closed) and G_n (open) such that

$$F_n \subset A_n \subset G_n, \qquad P(G_n - F_n) < 2^{-n}\varepsilon.$$

Then we have

$$\bigcup_n F_n \subset \bigcup_n A_n \subset \bigcup_n G_n,$$

$$P\left(\bigcup_n G_n - \bigcup_n F_n\right) \leq \sum_n P(G_n - F_n) \qquad \text{(Exercise 2.1 (iv))}$$

$$< \varepsilon.$$

By taking N sufficiently large, we obtain

$$P\left(\bigcup_n G_n - \bigcup_{n=1}^N F_n\right) < \varepsilon, \qquad \bigcup_{n=1}^N F_n \subset \bigcup_n A_n \subset \bigcup_n G_n.$$

Hence $\bigcup_n A_n \in \mathscr{B}$. Therefore $\mathscr{B}$ is closed under countable unions. Thus we have proved that $\mathscr{B}$ is a σ-algebra on S containing all open sets, which implies that $\mathscr{B} \supset \mathscr{B}(S)$.

By the discussion above we see that for every Borel set A and every $\varepsilon > 0$ there exists a closed set $F \supset A$ such that $P(A-F) < \varepsilon$. Since P is regular, this holds true for every P-measurable set A. Taking a compact set K with $P(S-K) < \varepsilon$, whose existence was proved above, we obtain

$$P(A - F \cap K) \leq P(A - F) + P(A \setminus K) < 2\varepsilon,$$

which proves the K-regularity of A. Therefore P is K-regular. ∎

Lemma 2.4.2 (Luzin's theorem). Let S be a complete separable metric space, T a separable metric space, P a regular probability measure on S, and $f: S \to T$ a P-measurable [i.e., measurable $\mathscr{D}(P)/\mathscr{B}(T)$] map. Then for every P-measurable set A and every $\varepsilon > 0$ we can find a compact set $K = K(A, \varepsilon) \subset A$ such that (i) $P(A - K) < \varepsilon$, and (ii) the restriction map $f_K \equiv f|_K : K \to T$ is continuous.

Proof: The case $S = T = \mathbb{R}^1$ is known as Luzin's theorem. We can prove our lemma essentially in the same way as this special case. Let d denote the metrics on T. Taking a sequence $\{b_n\}$ dense in T, we set

$$T = \bigcup_n B_{nk}, \qquad B_{nk} = \{ y \in T \mid d(y, b_n) \leq 1/k \}$$

and

$$A_{nk} = A \cap f^{-1}(B_{nk}), \qquad A'_{nk} = A_{nk} \setminus \bigcup_{i<n} A_{ik}.$$

Then A'_{nk} is P-measurable and

$$A = A \cap S = A \cap f^{-1}(T)$$
$$= \bigcup_n A \cap f^{-1}(B_{nk}) = \bigcup_n A_{nk} = \sum_n A'_{nk}.$$

Using the last lemma, we can find a compact set $K_{nk} \subset A'_{nk}$ such that

$$P(A'_{nk} - K_{nk}) \leq 2^{-n-k}\varepsilon.$$

Therefore

$$P\left(A - \sum_n K_{nk}\right) \leq \sum_n P(A'_{nk} - K_{nk}) \qquad \text{(Exercise 2.1 (iv))}$$

$$< 2^{-k}\varepsilon.$$

Taking $N(k)$ sufficiently large and setting $K_k = \sum_{n=1}^{N(k)} K_{nk}$ we have

$$P(A - K_k) < 2^{-k}\varepsilon,$$

where K_k is compact. Setting $K = \cap_{k=1}^{\infty} K_k$, we obtain

$$P(A - K) \leq \sum_k P(A - K_k) < \varepsilon,$$

where K is compact. To complete the proof it suffices to verify the continuity of $f_K = f|_K \colon K \to T$. It is obvious by $K \subset K_k$ that

$$K = K \cap K_k = \sum_{n=1}^{N(k)} K \cap K_{nk}.$$

If there is an empty set in the summands, we will remove it. Take a point a_{nk} in K_{nk} and define $g_k \colon K \to T$ by

$$g_k(x) = f(a_{nk}), \qquad x \in K_{nk}, n = 1, 2, \ldots, N(k)$$

$K \cap K_{nk}$ $(n = 1, 2, \ldots)$ and $N(k)$ are disjoint compact sets and g_k is constant on each set $K \cap K_{nk}$. Hence g_k is continuous. Since $f(a_{nk})$, $f(x)$ $\in f(K_{nk}) \subset f(A_{nk}) \subset B_{nk}$ for $x \in K \cap K_{nk}$, we obtain

$$d(g_k(x), f(x)) = d(f(a_{nk}), f(x)) \leq 2/k,$$

$$n = 1, 2, \ldots, N(k); x \in K \cap K_{nk},$$

so

$$d(g_k(x), f_K(x)) \leq 2/k \quad \text{for } x \in K.$$

Therefore $\{g_k\}$ converges to f_K uniformly. This implies the continuity of f_K. ∎

Let S and T be general sets, P a probability measure on S, and f an arbitrary map from S into T. Define a set function Q on T by

$$\mathscr{D}(Q) = \{ B \subset T | f^{-1}(B) \in \mathscr{D}(P) \}, \qquad Q(B) = P(f^{-1}(B)).$$

Q is clearly a probability measure on T, called the *image measure* of P under f, written Pf^{-1} or fP. If P is complete, then Q is also complete.

Lemma 2.4.3. Let S be a complete separable metric space, T a separable metric space, P a regular probability measure on S, and $f \colon S \to T$ a P-measurable map. Then the image measure $Q = Pf^{-1}$ is a K-regular probability measure on T.

Proof: It is obvious that Q is a complete probability measure on T whose domain includes $\mathscr{B}(T)$. To prove the regularity of Q it suffices to show that for every $B \in \mathscr{D}(Q)$ and every $\varepsilon > 0$ we can find a compact set $K \subset B$ such that $Q(B - K) < \varepsilon$. Since $B \in \mathscr{D}(Q)$, $A = f^{-1}(B)$ is P-measurable, so we can find a compact set $H \subset A$ such that $P(A - H) < \varepsilon$ and $f_H = f|_H \colon H \to T$ is continuous. Set $K = f_H(H) = f(H)$. Then K is compact and $f^{-1}(K) \supset H$. Hence

$$Q(B - K) = P(f^{-1}(B - K)) = P(f^{-1}(B) - f^{-1}(K)) \leq P(A - H) < \varepsilon.$$

∎

Now we are in a position to prove Theorem 2.4.1. Let B_n, $n = 1, 2, \ldots$ be a rearrangement of the sets B_{nk} ($n, k = 1, 2, \ldots$) introduced in the proof of Lemma 2.4.1 and let e_n be the indicator of B_n for $n = 1, 2, \ldots$. Define a real-valued function f on S by

$$f(x) = \sum_{n=1}^{\infty} \frac{2e_n(x)}{3^n}.$$

Denote Pf^{-1} by Q. Then Q is a regular probability measure on $\mathbb{R}^1$ by Lemma 2.4.3. It remains only to prove that (S, P) is isomorphic to $(\mathbb{R}^1, Q)$. If x and y are two different points in S, then $x \in B_n$ and $y \notin B_n$ for some n and so $f(x) \neq f(y)$. Hence f is 1-to-1. Define $f': S \to T \equiv f(S) \subset \mathbb{R}^1$ by restricting the range of f. Then f' is a bijective map. Since $f^{-1}(T) = S$, we have $Q(T) = 1$. Hence the restriction $Q' = Q|_T$ is a probability measure on T. Since $f(A) = f'(A) \subset T$ for $A \subset S$ and $Q = Pf^{-1}$, we obtain

$$A \in \mathscr{D}(P) \Leftrightarrow f'(A) \in \mathscr{D}(Q') \Rightarrow Q'(f'(A)) = P(A),$$

and so $(S, P) \approx (T, Q')$ (f'). This implies that (S, P) is isomorphic to $(\mathbb{R}^1, Q)$. ∎

Let us mention some examples of this theorem.

(a) A countable set Ω is a complete separable metric space with the metric

$$\rho(x, y) = \begin{cases} 1 & (x \neq y) \\ 0 & (x = y) \end{cases},$$

where $B(\Omega) = 2^{\Omega}$. Hence every probability measure on Ω defined for all subsets of Ω is standard.

(b) Let Ω be the product of a countable family of complete separable metric spaces $S_n = (S_n, \rho_n)$. We regard Ω as a topological space with the product topology. This topology is induced by the metric

$$\rho((x_n), (y_n)) = \sum_{n=1}^{\infty} 2^{-n} [\rho_n(x_n, y_n) \wedge 1], \qquad a \wedge b = \min(a, b).$$

As (S_n, ρ_n) is complete and separable for every n, so is (Ω, ρ). Hence every regular probability measure on Ω is standard by the theorem.

(c) The space $\mathbb{R}^{\infty} = \mathbb{R}^1 \times \mathbb{R}^1 \times \cdots$ is a complete separable metric space by fact (b) proved above. Since $\mathbb{R}^{\infty}$ is the space of all sequences of real numbers, it is often called the *sequence space*. We denote the Borel system $\mathscr{B}(\mathbb{R}^{\infty})$ by $\mathscr{B}^{\infty}$. Now we will prove that

$$\mathscr{B}^{\infty} = \mathscr{B}^1 \times \mathscr{B}^2 \times \cdots.$$

Note that this is not obvious, because the right-hand side is the product σ-algebra defined in the last section. Since the n-projection π_n: $(x_1, x_2, \ldots) \mapsto x_n$ is continuous, $\pi_n^{-1}(E)$ belongs to $\mathscr{B}^\infty$ for every $E \in \mathscr{B}^1$. This implies that

$$\mathscr{B}^\infty \supset \mathscr{B}^1 \times \mathscr{B}^1 \times \cdots .$$

The opposite inclusion relation follows from the fact that every open subset of $\mathbb{R}^\infty$ (with respect to the product topology) is represented by a countable union of sets of the form:

$$(a_1, b_1) \times (a_2, b_2) \times \cdots \times (a_n, b_n) \times \mathbb{R}^1 \times \mathbb{R}^1 \times \cdots .$$

Since $\mathbb{R}^1$ is a complete separable metric space, every regular probability measure on $\mathbb{R}^\infty$ is standard by the fact (b).

Theorem 2.4.2. Let P_n be a regular probability measure on a complete separable metric space S_n for $n = 1, 2, \ldots$. Then the complete direct product $P = \overline{\prod_n P_n}$ is a regular (and so standard) probability measure on the product space $S = \prod_n S_n$.

Proof: Let μ_n denote the restriction of P_n to $\mathscr{B}(S_n)$. Then the product measure $\mu = \prod_n \mu_n$ is a probability measure on $\prod_n \mathscr{B}(S_n)$. This σ-algebra coincides with the Borel system $\mathscr{B}(S)$ with respect to the product topology, as we can prove by the argument used in (b). Hence μ is a Borel probability measure on S. Since

$$P = \overline{\prod_n P_n} = \overline{\prod_n \mu_n}$$
$$= \overline{\prod_n \bar{\mu}_n} \qquad \text{(Exercise 2.3 (ii))}$$
$$= \bar{\mu},$$

P is a regular probability measure on S. ∎

Theorem 2.4.3. The complete direct product of a countable family of standard probability measures is standard.

Proof: This follows easily from the following facts:
(a) If $(\Omega_n, P_n) \approx (\Omega_n', P_n')\ (f_n),\ n = 1, 2, \ldots$, then

$$\left(\prod_n \Omega_n, \prod_n P_n \right) \approx \left(\prod_n \Omega_n', \prod_n P_n' \right) (f),$$

where $f(\omega_1, \omega_2, \ldots) = (f_1(\omega_1), f_2(\omega_2), \ldots)$.

(b) If $P_n(A_n) = 1$ $(n = 1, 2, \ldots)$ and $P = \prod_n P_n$, then

$$P\left(\prod_n A_n\right) = P\left(\lim_{n \to \infty} A_1 \times A_2 \times \cdots \times A_n \times \Omega_{n+1} \times \Omega_{n+2} \times \cdots\right)$$

$$= \lim_{n \to \infty} P(A_1 \times A_2 \times \cdots \times A_n \times \Omega_{n+1} \times \Omega_{n+2} \times \cdots)$$

$$= \lim_{n \to \infty} P_1(A_1) P_2(A_2) \ldots P_n(A_n) = 1. \qquad \blacksquare$$

Exercise 2.4. Let T be a Borel subset of a complete separable metric space $S = (S, \rho)$. Then T is a separable metric space with the restriction $\rho_T = \rho|_T$. Prove that every regular probability measure P on T is standard.

[*Hint:* Setting $Q(B) = P(B \cap T)$, we obtain a regular probability measure on S. Since $Q(T) = 1$, we obtain $P = Q|_T$, so $(T, P) \sim (S, Q)$.]

2.5 One-dimensional distributions

Let μ be a regular probability measure on $\mathbb{R}^1$. Then the restriction of μ to $\mathcal{B}^1$, $\mu|_{\mathcal{B}^1}$, is a Borel probability measure on $\mathbb{R}^1$, and μ is the Lebesgue extension of $\mu|_{\mathcal{B}^1}$. Hence every regular probability measure μ is completely determined by its behavior on $\mathcal{B}^1$. Also, μ is determined by its distribution function:

$$F(x) = \mu(-\infty, \mu],$$

as we mentioned in Section 2.2.

From now on we will call a regular probability measure on $\mathbb{R}$ simply a *one-dimensional distribution* or a *distribution*.

2.5a Lebesgue decomposition

Let $\mu, \nu, \ldots$ represent distributions. A point $a \in \mathbb{R}^1$ is called a *discontinuity point* of μ if

$$\mu\{a\} > 0,$$

namely, if the distribution function of μ has a jump at a. The set of all discontinuity points of μ, denoted by D_μ, is a countable set. A point $a \in \mathbb{R}^1$ is called a *continuity point* of μ if $\mu\{a\} = 0$. The set of all continuity points of μ is denoted by C_μ. μ is called *purely discontinuous* if $\mu(D_\mu) = 1$. Let μ be purely discontinuous and $D_\mu = \{a_n | n = 1, 2, \ldots\}$. Then

$$\mu(E) = \sum_{a_n \in E} \mu\{a_n\}.$$

Hence μ is determined by $\{a_n\}$ with $\{p_n = \mu(a_n)\}$, and so is represented by the form

$$\begin{pmatrix} a_1 & a_2 & \cdots \\ p_1 & p_2 & \cdots \end{pmatrix} \quad \text{or} \quad \begin{pmatrix} a_n \\ p_n \end{pmatrix}_{n=1,2,\ldots},$$

where

$$p_n \geqq 0, \quad \sum_n p_n = 1,$$

Let us mention some purely discontinuous distributions:

δ-distribution

$$\delta_a = \begin{pmatrix} a \\ 1 \end{pmatrix} \quad (a \in \mathbb{R}^1).$$

δ_0 is simply denoted by δ.

Binomial distribution

$$b_{n,p} = \left(\begin{matrix} k \\ \binom{n}{k} p^k (1-p)^{n-k} \end{matrix} \right)_{k=0,1,2,\ldots,n}$$

$$(n = 1, 2, \ldots, 0 < p < 1).$$

Poisson distribution

$$p_\lambda = \left(\begin{matrix} k \\ e^{-\lambda} \lambda^k / k! \end{matrix} \right)_{k=0,1,2,\ldots} \quad (\lambda > 0).$$

If μ has no discontinuity point, then μ is called a *continuous distribution*. If μ is continuous, then $\mu\{a\} = 0$ for every $a \in \mathbb{R}^1$ and so $\mu(C) = 0$ for every countable set C. μ is called *absolutely continuous* if $\mu(E) = 0$ whenever $E \in \mathscr{B}^1$ and $\lambda(E) = 0$, where $\lambda(E)$ denotes the Lebesgue measure of E. Using the Radon–Nikodym theorem in measure theory, we can see that μ is absolutely continuous if and only if it has density; that is,

$$\mu(E) = \int_E f(x)\, dx \quad (E \in \mathscr{B}^1)$$

or

$$\mu(dx) = f(x)\, dx,$$

where f is Lebesgue-measurable and

$$f(x) \geqq 0, \quad \int_{\mathbb{R}^1} f(x)\, dx = 1.$$

The Gauss distribution and the Cauchy distribution (Exercise 2.1(i)) are absolutely continuous.

Absolute continuity implies continuity, but its converse is not true. In fact there exists a continuous distribution μ such that

$$\mu(E) = 1 \quad \text{and} \quad \lambda(E) = 0$$

for some $E \in \mathscr{B}^1$. Such a distribution is called *singular*.

We will construct the Cantor distribution, which is a typical example of a singular distribution. Denote the Lebesgue probability measure on $[0, 1]$ by P and define a function $f: [0, 1] \to \mathbb{R}^1$ by

$$x = \sum_{n=1}^{\infty} \frac{\varepsilon_n}{2^n} \mapsto f(x) = \sum_{n=1}^{\infty} \frac{2\varepsilon_n}{3^n},$$

where $\varepsilon_n \ (= 0 \text{ or } 1)$ is given as follows:

$$\varepsilon_n(x) = [2^n x] - 2[2^{n-1} x]$$

([ ] here is the Gauss bracket.) As $\varepsilon_n(x)$, $n = 1, 2, \dots$ are Borel-measurable in x, so is $f(x)$. Hence the image measure $\mu = Pf^{-1}$ is a regular probability measure on $\mathbb{R}^1$. Since $f([0, 1])$ is included in the Cantor set $\mathbf{K}$ and since f is strictly increasing, we have

$$\mu(\mathbf{K}) = 1 \quad \text{and} \quad \mu\{a\} = 0 \quad (a \in \mathbb{R}^1).$$

Since the Lebesgue measure of $\mathbf{K}$ is 0, μ is a singular distribution.

If $\mu_1, \mu_2, \dots$ are distributions, then

$$\mu(E) = \sum_n c_n \mu_n(E), \qquad E \in \mathscr{B}^1; c_n \geq 0, \sum_n c_n = 1,$$

is also a distribution. This is called a *convex combination* of $\{\mu_n\}$.

Theorem 2.5.1 (The Lebesgue decomposition theorem). Every distribution is uniquely represented as a convex combination of a purely discontinuous distribution, an absolutely continuous distribution, and a singular distribution.

Proof: Since the uniqueness of the decomposition is easy to prove, we will prove the existence only.

Let μ be an arbitrary distribution. If we define two measures ν_d and ν_c by

$$\nu_d(E) = \mu(E \cap D_\mu), \qquad \nu_c(E) = \mu(E \cap C_\mu)$$

then we have

$$\mu = \nu_d(\mathbb{R}^1) \frac{\nu_d}{\nu_d(\mathbb{R}^1)} + \nu_c(\mathbb{R}^1) \frac{\nu_c}{\nu_c(\mathbb{R}^1)},$$

where the term with coefficient 0 is deleted. This expression shows that every distribution is a convex combination of a purely discontinuous distribution and a continuous distribution. Hence is suffices to prove that every continuous distribution μ is a convex combination of a singular distribution and an absolutely continuous distribution. If we set

$$s = \sup\{\mu(E)|E \in \mathscr{B}^1, \lambda(E) = 0\},$$

then we can find a sequence of Borel sets $\{E_n\}$ such that

$$\lambda(E_n) = 0, \qquad \mu(E_n) \to s.$$

Then the union $S = \bigcup_n E_n$ satisfies

$$\lambda(S) = 0, \qquad \mu(S) = s.$$

If $A \subset S^c$, $\lambda(A) = 0$, and $\mu(A) > 0$, then

$$\lambda(S + A) = \lambda(S) = 0 \qquad \text{and} \qquad \mu(S + A) > \mu(S) = s.$$

This contradicts the definition of s. Hence

$$A \subset S^c, \qquad \lambda(A) = 0 \Rightarrow \mu(A) = 0.$$

Keeping this in mind and setting

$$\nu_s(E) = \mu(E \cap S), \qquad \nu_{ac}(E) = \mu(E \cap S^c), \qquad E \in \mathscr{B}^1,$$

we obtain

$$\nu_s(S) = \mu(S), \qquad \lambda(S) = 0$$

and

$$\lambda(E) = 0 \Rightarrow \lambda(E \cap S^c) = 0 \Rightarrow \nu_{ac}(E) = \mu(E \cap S^c) = 0.$$

Hence

$$\mu = \nu_s(\mathbb{R}^1) \frac{\nu_s}{\nu_s(\mathbb{R}^1)} + \nu_{ac}(\mathbb{R}^1) \frac{\nu_{ac}}{\nu_{ac}(\mathbb{R}^1)}$$

is the decomposition that was to be sought. ∎

2.5b *Convergence of distributions*

A sequence of distributions $\{\mu_n\}$ is said to *converge* to a distribution μ (written $\mu_n \to \mu$) if

$$\int_{\mathbb{R}^1} f(x)\mu_n(dx) \to \int_{\mathbb{R}^1} f(x)\mu(dx) \qquad (n \to \infty)$$

for every bounded continuous function f on $\mathbb{R}^1$.

It is easy to prove that if

$$\mu_n(E) \to \mu(E) \qquad (n \to \infty)$$

for every Borel set E, then $\mu_n \to \mu$. But its converse is not always true, as we can see in the following examples:

(i) $a_n \to a$ implies $\delta_{a_n} \to \delta_a$, but

$$\delta_{a_n}(E) = 0 \qquad (a_n \neq a),$$

$$\delta_a(E) = 1$$

for $E = \{a\}$.

(ii) $v_n \to 0$ implies $N_{0, v_n} \to \delta$, but because we have

$$\int_{\mathbf{R}^1} f(x) N_{0, v_n}(dx) = \int_{\mathbf{R}^1} f(x) \frac{1}{\sqrt{2\pi v_n}} e^{-x^2/2v_n} \, dx$$

$$= \int_{\mathbf{R}^1} f(\sqrt{v_n} \, y) \frac{1}{\sqrt{2\pi}} e^{-y^2/2} \, dy$$

$$\to \int_{\mathbf{R}^1} f(0) \frac{1}{\sqrt{2\pi}} e^{-y^2/2} \, dy$$

$$= f(0) = \int_{\mathbf{R}^1} f(x) \delta(dx)$$

for every bounded continuous function f. But we have

$$N_{0, v_n}(E) = \tfrac{1}{2}, \qquad \delta(E) = 1 \qquad (E = [0, \infty)).$$

The following theorem gives necessary and sufficient conditions for the convergence of distributions.

Theorem 2.5.2. The following conditions are equivalent to each other, where μ_n and μ are distributions and F_n and F are their distribution functions, respectively:

(i) $\mu_n \to \mu$.

(ii) $\int_{\mathbf{R}^1} g(x) \mu_n(dx) \to \int_{\mathbf{R}^1} g(x) \mu(dx)$ for every continuous function g with compact support, where the support of g is the closure of the set $\{x \mid g(x) \neq 0\}$.

(iii) $\mu_n(E) \to \mu(E)$ for every Borel set E with $\mu(E^0) = \mu(\bar{E})$, E^0 and $\bar{E}$ denoting the open kernel and the closure of E, respectively.

(iv) $F_n(x) \to F(x)$ for every continuity point x of F.

(v) There exists a countable dense subset C of $\mathbf{R}^1$ such that $F_n(x) \to F(x)$ for every $x \in C$.

Proof: Since it is obvious that (iii) $\Rightarrow$ (iv) $\Rightarrow$ (v), it suffices to prove that (v) $\Rightarrow$ (ii) $\Rightarrow$ (i) $\Rightarrow$ (iii).

$(v) \Rightarrow (ii)$ For every continuous function g with compact support we can find a left-continuous step function g_ε with compact support such that

$$\sup_x |g_\varepsilon(x) - g(x)| < \varepsilon.$$

Since the set C in assumption (v) is dense in $\mathbb{R}^1$, we may assume that the jumping points of g_ε, $a_i = a_i(\varepsilon)$ $(i = 1, 2,, \ldots, m = m(\varepsilon))$ belong to C. We will denote the integral

$$\int_{\mathbb{R}^1} g(x)\nu(dx)$$

by (g, ν). Then we have

$$|(g, \mu_n) - (g, \mu)| \leq |(g, \mu_n) - (g_\varepsilon, \mu_n)| + |(g, \mu) - (g_\varepsilon, \mu)|$$

$$+ |(g_\varepsilon, \mu_n) - (g_\varepsilon, \mu)|$$

$$\leq \varepsilon + \varepsilon + \left| \sum_{i=1}^{m} g_\varepsilon(a_i)(F_n(a_i) - F_n(a_{i-1})) \right.$$

$$\left. - \sum_{i=1}^{m} g_\varepsilon(a_i)(F(a_i) - F(a_{i-1})) \right|.$$

Since $\lim_{n \to \infty} F_n(a_i) = F(a_i)$, the third term of the last expression tends to 0 as $n \to \infty$. Hence

$$\limsup_{n \to \infty} |(g, \mu_n) - (g, \mu)| \leq 2\varepsilon.$$

Since ε is any positive number, the left-hand side must vanish. This proves that (v) implies (ii).

$(ii) \Rightarrow (i)$. Let f be any bounded continuous function. We will prove that $(f, \mu_n) \to (f, \mu)$ under assumption (ii). For every $m > 0$ we can find a continuous function g that coincides with f on $[-m, m]$ and vanishes on $[-m - 1, m + 1]^c$. Now let us observe the inequality:

$$|(f, \mu_n) - (f, \mu)| \leq (|f - g|, \mu_n) + (|f - g|, \mu) + |(g, \mu_n) - (g, \mu)|.$$

Since $f - g$ is bounded, we have a constant c such that $|f(x) - g(x)| \leq c$. Take a continuous function $h(x)$ that vanishes on $[-m + 1, m - 1]$, equals c on $[-m, m]^c$ and lies between 0 and c elsewhere. Since $|f(x) - g(x)|$ vanishes on $[-m, m]$ and is bounded by c, it is bounded by c, it is obvious that

$$|f(x) - g(x)| \leq h(x),$$

so

$$(|f - g|, \mu_n) \leq (h, \mu_n) = (c, \mu_n) - (c - h, \mu_n) = c - (c - h, \mu_n).$$

Since $c - h$ is continuous and has compact support $\subset [-m, m]$, we have

$$(c - h, \mu_n) \to (c - h, \mu) = c - (h, \mu)$$

by assumption (ii). This implies that

$$\limsup_{n \to \infty} (|f - g|, \mu_n) \leq (h, \mu) \leq c\mu([-m + 1, m - 1]^c).$$

Also we have

$$(|f - g|, \mu) \leq (h, \mu) \leq c\mu([-m + 1, m - 1]^c).$$

Since g is continuous and has compact support, we have

$$\lim_{n \to \infty} |(g, \mu_n) - (g, \mu)| = 0.$$

Thus, we obtain

$$\limsup_{n \to \infty} |(f, \mu_n) - (f, \mu)| \leq 2c\mu([-m + 1, m - 1]^c).$$

Since the right-hand side tends to 0 as $m \to \infty$, the left-hand side must vanish.

$(i) \to (iii)$. Suppose that E has the property mentioned in assumption (iii). Every open set in $\mathbf{R}^1$ is the limit of an increasing sequence of closed sets and every closed set in $\mathbf{R}^1$ is the limit of a decreasing sequence of open sets. Hence we can find a closed set $F_\varepsilon \subset E^0$ and an open set $G_\varepsilon \supset \bar{E}$ such that

$$\mu(E^0) - \mu(F_\varepsilon) < \varepsilon, \qquad \mu(G_\varepsilon) - \mu(\bar{E}) < \varepsilon,$$

where ε is an arbitrary positive number. This implies

$$\mu(G_\varepsilon) < \mu(E) + \varepsilon, \qquad \mu(F_\varepsilon) > \mu(E) - \varepsilon$$

by the property of E. Let f_ε be a continuous function that vanishes on G_ε^c, equals 1 on $\bar{E}$, and lies between 0 and 1 elsewhere. Then

$$\lim_{n \to \infty} (f_\varepsilon, \mu_n) = (f_\varepsilon, \mu)$$

by assumption (i). Also, it is obvious that

$$\mu_n(E) \leq \mu_n(\bar{E}) \leq (f_\varepsilon, \mu_n).$$

Hence we obtain

$$\limsup_{n \to \infty} \mu_n(E) \leq (f_\varepsilon, \mu) \leq \mu(G_\varepsilon) \leq \mu(E) + \varepsilon,$$

which proves that

$$\limsup_{n \to \infty} \mu_n(E) \leq \mu(E).$$

Similarly, we can prove that

$$\liminf_{n \to \infty} \mu_n(E) \geq \mu(E).$$

Therefore, we can see that $\lim_{n \to \infty} \mu_n(E) = \mu(E)$. ■

The limit of a convergent sequence of distributions $\{\mu_n\}$ is unique. Suppose that $\mu_n \to \mu$ and $\nu_n \to \mu$ and denote the distribution functions of μ and ν by F and G, respectively. Then $F = G$ on $(D_\mu \cup D_\nu)^c$ by the theorem above. Since $D_\mu \cup D_\nu$ is countable and F and G are right-continuous, $F = G$ everywhere, namely, $\mu = \nu$. Also, it is obvious that if $\mu_n \to \mu$, then every subsequence of $\{\mu_n\}$ converges to μ.

Theorem 2.5.3. Let $\mathcal{M}$ be a set of distributions. Then the following conditions are equivalent to each other:

  (i) Every sequence of distributions in $\mathcal{M}$ has a convergent subsequence. (The limit distribution may not belong to $\mathcal{M}$.)
  (ii) $\lim_{a \to \infty} \inf_{\mu \in \mathcal{M}} \mu[-a, a] = 1$.

Proof: Suppose that (i) holds and that (ii) does not hold. Then we can find a positive number ε_0 and a sequence $\{\nu_n\}$ in $\mathcal{M}$ such that

$$\nu_n[-n, n] \leq 1 - \varepsilon_0, \qquad n = 1, 2, \dots.$$

By assumption (i) we can find a convergent subsequence $\{\nu_{n(k)}\}$ of $\{\nu_n\}$. Denote the limit of $\{\nu_{n(k)}\}$ by ν. Take $\alpha_k \in (-n(k), -n(k)+1) \cap C_\nu$ and $\beta_k \in (n(k)-1, n(k)) \cap C_\nu$. If $l \geq k$, then

$$\nu_{n(l)}(\alpha_k, \beta_k) \leq \nu_{n(l)}[-n(k), n(k)] \leq \nu_{n(l)}[-n(l), n(l)]$$

$$\leq 1 - \varepsilon_0.$$

Letting $l \to \infty$ and then letting $k \to \infty$, we obtain

$$\nu(-\infty, \infty) \leq 1 - \varepsilon_0,$$

in contradiction with $\nu(-\infty, \infty) = 1$. Therefore, (i) implies (ii).

Next we will find a convergent subsequence of $\{\mu_n\}$ under assumption (ii). Denote the distribution function of μ_n by F_n. Take a sequence $\{a_n\}$ dense in $\mathbb{R}^1$. Then

$$0 \leq F_n(a_m) \leq 1, \qquad n, m = 1, 2, \dots.$$

Hence we can find a subsequence of $\{F_n\}$, $\{F_{n(m,k)}, k=1,2,\ldots\}$ for every m such that

$$F_{n(1,k)}(a_1) \to \tilde{F}(a_1)$$

$$F_{n(2,k)}(a_2) \to \tilde{F}(a_2)$$

$$\vdots$$

Also, we can determine $\{F_{n(m,k)}, k=1,2,\ldots\}$ so that $\{F_{n(m+1,k)}, k=1,2,\ldots\}$ is a subsequence of $\{F_{n(m,k)}, k=1,2,\ldots\}$ for every m. Then $\{F_{n(k,k)}, k=m,m+1,\ldots\}$ is a subsequence of $\{F_{n(m,k)}, k=1,2,\ldots\}$. Hence

$$\lim_{k\to\infty} F_{n(k,k)}(a_m) = \tilde{F}(a_m).$$

It is obvious that

$$0 \leqq \tilde{F}(a_m) \leqq 1,$$

$$\tilde{F}(a_m) \leqq \tilde{F}(a_l) \qquad (a_m \leqq a_l).$$

Since $\{a_n\}$ is dense in $\mathbb{R}^1$, we can use assumption (ii) to find $a_m = a_m(\varepsilon)$ and $a_l = a_l(\varepsilon)$ such that

$$F_n(a_m) - F_n(a_l) > 1 - \varepsilon, \qquad n=1,2,\ldots,$$

where ε is an arbitrary positive number. Then

$$\tilde{F}(\infty) - \tilde{F}(-\infty) \geqq \tilde{F}(a_m) - \tilde{F}(a_l) \geqq 1 - \varepsilon,$$

which implies that $\tilde{F}(\infty) = 1$ and $\tilde{F}(-\infty) = 0$. Define $F(x)$ by

$$F(x) = \inf_{a_m > x} \tilde{F}(a_m).$$

Then $F(x)$ is a right-continuous increasing function with $F(-\infty) = 0$ and $F(\infty) = 1$ and so $F(x)$ is the distribution function of a distribution ν. It remains only to prove that $\nu_n(k,k) \to \nu$ as $k \to \infty$. Denote the distribution function of $\nu_{n(k,k)}$ by G_k and let a be an continuity point of F. Take $\delta = \delta(\varepsilon) > 0$ for every $\varepsilon > 0$ such that

$$F(a) - \varepsilon < F(a-\delta).$$

Since $\{a_m\}$ is dense in $\mathbb{R}^1$, we can find $a_m \in (a-\delta, a)$. Then

$$F(a) - \varepsilon < F(a-\delta) \leqq \tilde{F}(a_m) = \lim_{k\to\infty} G_k(a_m) \leqq \liminf_{k\to\infty} G_k(a).$$

Also, we can find $a_l > a$ such that

$$F(a) + \varepsilon > \tilde{F}(a_l) = \lim_{k\to\infty} G_k(a_l) \geqq \limsup_{k\to\infty} G_k(a).$$

Thus we have

$$F(a) \leq \liminf_{k \to \infty} G_k(a), \qquad F(a) \geq \limsup_{k \to \infty} G_k(a),$$

so $F(a) = \lim_{k \to \infty} G_k(a)$. This completes the proof. ∎

2.5.c   Moments

The pth-order *moment* of a distribution μ, $M_p(\mu)$ ($p = 1, 2, \ldots$) is defined by

$$M_p(\mu) = \int_{\mathbb{R}} x^p \mu(dx).$$

Also the pth-order *absolute moment* $|M|_p(\mu)$ ($0 < p < \infty$) is defined by

$$|M|_p(\mu) = \int_{\mathbb{R}} |x|^p \mu(dx) \qquad (\in [0, \infty]).$$

$M_p(\mu)$ ($p = 1, 2, \ldots$) is well-defined and finite if $|M|_p(\mu) < \infty$.

Lemma 2.5 (Hölder's inequality). If $p, q \in (1, \infty)$ and $p^{-1} + q^{-1} = 1$, then

$$\int_{\mathbb{R}^1} |f(x) g(x)| \mu(dx) \leq \left(\int_{\mathbb{R}^1} |f(x)|^p \mu(dx) \right)^{1/p} \left(\int_{\mathbb{R}^1} |g(x)|^q \mu(dx) \right)^{1/q}.$$

Proof: Let a and b denote the two factors of the right-hand side. If either a or b is 0 or ∞, the inequality is obvious. (We put $0 \cdot \infty = \infty \cdot 0 = 0$ for convention.) Hence we may assume that both a and b are positive and finite. By replacing f and g by f/a and g/b, respectively, we can see that it suffices to observe the case where $a = b = 1$. The function

$$F(x) = -\log x, \qquad 0 < x < \infty,$$

is convex because

$$F''(x) = 1/x^2 > 0.$$

Therefore

$$-\frac{1}{p} \log \alpha^p - \frac{1}{q} \log \beta^q \geq -\log \left(\frac{1}{p} \alpha^p + \frac{1}{q} \beta^q \right), \qquad \alpha, \beta > 0,$$

namely,

$$\alpha\beta \leq \frac{1}{p} \alpha^p + \frac{1}{q} \beta^q, \qquad \alpha, \beta > 0.$$

Also, this inequality holds even if either α or β vanishes. Hence

$$|f(x) g(x)| \leq \frac{1}{p} |f(x)|^p + \frac{1}{q} |g(x)|^q$$

holds for every $x \in \mathbb{R}^1$. Now integrate both sides over $\mathbb{R}^1$ to complete the proof. ∎

Theorem 2.5.4. $1 \leq p < q < \infty \Rightarrow (|M|_p(\mu))^{1/p} \leq (|M|_q(\mu))^{1/q}$.

Proof: Putting $r = q/p$ and $s = q/(q-p)$ we have

$$1/r + 1/s = 1.$$

Using Hölder's inequality we have

$$\int_{\mathbb{R}^1} |x|^p \mu(dx) \leq \left(\int_{\mathbb{R}^1} (|x|^p)^r \mu(dx) \right)^{1/r} \left(\int_{\mathbb{R}^1} 1^s \mu(dx) \right)^{1/s}$$

$$= \left(\int_{\mathbb{R}^1} |x|^q \mu(dx) \right)^{p/q}. \qquad \blacksquare$$

Theorem 2.5.5 (Bienaymé's inequality).

$$\mu([-a, a]) \geq 1 - \frac{|M|_p(\mu)}{a^p}.$$

Proof:

$$|M|_p(\mu) = \int_{\mathbb{R}^1} |x|^p \mu(dx) \geq \int_{[-a, a]^c} |x|^p \mu(dx) \geq a^p \mu([-a, a]^c),$$

$$= a^p (1 - \mu[-a, a]) \qquad \blacksquare$$

$M_1(\mu)$ is called the mean of μ and

$$V(\mu) = \int_{\mathbb{R}^1} (x - M(\mu))^2 \mu(dx)$$

is called the variance of μ. The Gauss distribution $N_{m, v}$ has mean value m and variance v.

2.5d   Convolution

If μ_1 and μ_2 are distributions, then

$$\mu(E) = \int_{\mathbb{R}^1} \mu_1(E - x) \mu_2(dx), \qquad (E \in \mathscr{B}^1, E - x = \{ y - x | y \in E \})$$

is also a distribution and is called the *convolution* of μ_1 and μ_2. We will check that the convolution is always well-defined. Since μ is a distribution, the class of sets

$$\mathscr{A} = \{ E \in \mathscr{B}^1 | \mu_1(E - x) \text{ is Borel-measurable in } x \}$$

is a Dynkin class. Since

$$\mu_1((a, b] - x) = \mu_1(a - x, b - x]$$

is Borel-measurable in x, being left-continuous, $\mathscr{A}$ contains $(a, b]$ ($-\infty < a < b < \infty$) and so contains all Borel sets by the Dynkin class theorem. Hence the integral above is well-defined. It is easy to see that the μ defined above is a distribution. The convolution $\mu_1 * \mu_2$ can be represented as follows by Fubini's theorem on direct product measures:

$$(\mu_1 * \mu_2)(E) = \int_{\mathbb{R}^2} 1_E(x + y) \mu_1(dx) \mu_2(dy).$$

For every Borel-measurable function $f: \mathbb{R}^1 \to \mathbb{R}^1$ $f(x + y)$ is Borel-measurable in $(x, y) \in \mathbb{R}^2$, because $h: (x, y) \to f(x + y)$ is the composition of a continuous (so Borel-measurable) function $g: (x, y) \to x + y$ and a Borel-measurable function f. Since every bounded Borel-measurable function is represented as the limit of linear combinations of 1_E, $E \in \mathscr{B}^1$, we can deduce from the equality above that

$$\int_{\mathbb{R}^1} f(x) \mu(dx) = \iint_{\mathbb{R}^2} f(x + y) \mu_1(dx) \mu_2(dy)$$

holds for every bounded Borel-measurable function f. Due to this observation we can easily see that

$$\mu_1 * \mu_2 = \mu_2 * \mu_1, \qquad (\mu_1 * \mu_2) * \mu_3 = \mu_1 * (\mu_2 * \mu_3),$$

namely, that convolution is subject to commutative law and associative law. Hence we may write $\mu_1 * \mu_2 * \cdots * \mu_n$ omitting the brackets that indicate the order of convolutions. It is easy to verify that $\mu = \mu_1 * \mu_2 * \cdots * \mu_n$ if and only if

$$\int_{\mathbb{R}^1} f(x) \mu(dx) = \iint \cdots \int_{\mathbb{R}^n} f(x_1 + x_2 + \cdots + x_n) \mu_1(dx_1)$$
$$\times \mu_2(dx_2) \ldots \mu_n(dx_n)$$

for every bounded Borel-measurable function f.

Exercise 2.5.

(i) Prove that the following functions are densities of distributions and make graphs of their distribution functions ($-\infty < a < b <$

∞, $c > 0, \lambda > 0$):

$$U_{a,b}(x) = \begin{cases} (b-a)^{-1} & a < x < b \\ 0 & \text{elsewhere} \end{cases} \qquad \text{(uniform distribution)};$$

$$T_{a,c}(x) = \begin{cases} \dfrac{c - |a - x|}{c^2} & a - c < x < a + c \\ 0 & \text{elsewhere} \end{cases} \qquad \text{(triangular distribution)};$$

$$E_\lambda(x) = \begin{cases} \lambda e^{-\lambda x} & x \geq 0 \\ 0 & \text{elsewhere} \end{cases} \qquad \text{(exponential distribution)}.$$

(ii) Use Theorem 2.5.2$((i) \Leftrightarrow (iv))$ to prove the following:

$$\begin{pmatrix} k/n \\ 1/n \end{pmatrix}_{k=1,2,\ldots,n} \to U_{0,1} \qquad (n \to \infty),$$

$$\begin{pmatrix} k/n \\ (\lambda/n)(1 - \lambda/n)^{k-1} \end{pmatrix}_{k=1,2,\ldots} \to E_\lambda \qquad (n \to \infty),$$

$$U_{-1/n,1/n} \to \delta \qquad (n \to \infty).$$

(iii) Prove that if $\sup_{\mu \in \mathscr{M}} |M|_p(\mu) < \infty$, then every sequence of distributions in $\mathscr{M}$ has a convergent subsequence. [*Hint*: Use Theorems 2.5.3 and 2.5.5.]

(iv) Prove that if μ_1 has density f_1, then $\mu_1 * \mu_2$ has density $f(x) = \int_{\mathbb{R}^1} f_1(x - y)\mu_2(dy)$ and that if μ_i has density f_i for $i = 1, 2$, then $\mu_1 * \mu_2$ has density $f(x) = \int_{\mathbb{R}^1} f_1(x - y)f_2(y)\,dy$.

(v) Make the graph of the density of $U_{-c,c} * U_{a,b}$.

(vi) Prove that

$$N_{m,1/n} \to \delta_m, \qquad C_{m,1/n} \to \delta_m.$$

Remark: This fact justifies defining $N_{m,0}$ and $C_{m,0}$ to be δ_m.

(vii) Prove that $N_{m_1,v_1} * N_{m_2,v_2} = N_{m_1+m_2,v_1+v_2}$.

(viii) Prove that $C_{m_1,c_1} * C_{m_2,c_2} = C_{m_1+m_2,c_1+c_2}$. [*Hint*: Use the residue theorem of complex analysis to compute

$$\int_{-\infty}^{\infty} \frac{dy}{\left((x-y-m_1)^2 + c_1^2\right)\left((y-m_2)^2 + c_2^2\right)}.$$

(Observe the integral along $-a \to a \to a + ia \to -a + ia \to -a$ and let $a \to \infty$.)]

(ix) Prove that if

$$r_n = \sum_{k=0}^{n} p_k q_{n-k},$$

then

$$\begin{pmatrix} n \\ p_n \end{pmatrix}_{n=0,1,\ldots} * \begin{pmatrix} n \\ q_n \end{pmatrix}_{n=0,1,\ldots} = \begin{pmatrix} n \\ r_n \end{pmatrix}_{n=0,1,\ldots},$$

and use this fact to prove the following:

$$b_{n,p} * b_{m,p} = b_{n+m,p}, \qquad P_\lambda * P_\mu = P_{\lambda+\mu}.$$

2.6 Characteristic functions

2.6a Definition and examples
Let μ be a distribution. The function

$$\varphi(z) = \int_{\mathbb{R}^1} e^{izx} \mu(dx), \qquad z \in \mathbb{R}^1$$

is called the *characteristic function* of μ or the *Fourier transform* of μ, denoted by $\mathscr{F}\mu$. A function $g(z)$, $z \in \mathbb{R}^1$ is called a characteristic function if there exists a distribution μ such that $g = \mathscr{F}\mu$.

If μ is a purely discontinuous distribution of the form

$$\mu = \begin{pmatrix} a_1 & a_2 & \cdots \\ p_1 & p_2 & \cdots \end{pmatrix},$$

then

$$\mathscr{F}\mu(z) = \sum_n p_n e^{ia_n z}.$$

Hence the characteristic function of the Poisson distribution p_λ is

$$\sum_{n=0}^{\infty} e^{-\lambda} \frac{\lambda^n}{n!} e^{izn} = e^{-\lambda} \sum_{n=0}^{\infty} \frac{(\lambda e^{iz})^n}{n!} = e^{-\lambda} e^{\lambda e^{iz}} = \exp\{\lambda(e^{iz} - 1)\}.$$

If μ has density f, then

$$\mathscr{F}\mu(z) = \int_{\mathbb{R}^1} e^{izx} f(x)\, dx.$$

For example, the characteristic function of the uniform distribution $U_{a,b}$

is

$$\frac{1}{b-a}\int_a^b e^{izx}\,dx = \frac{e^{izb}-e^{iza}}{iz(b-a)} = e^{iz(a+b)/2}\frac{e^{iz(b-a)/2}-e^{-iz(b-a)/2}}{2iz(b-a)/2}$$

$$= e^{iz(a+b)/2}\frac{\sin z(b-a)/2}{z(b-a)/2}.$$

Let us prove that the characteristic function of the Gauss distribution $N_{m,\,v}$ is

$$\int_{-\infty}^{\infty} e^{izx}\frac{1}{\sqrt{2\pi v}}e^{-(x-m)^2/2v}\,dx = \exp\left\{imz - \frac{v}{2}z^2\right\} \qquad (v>0).$$

Note that this fact holds obviously even in case $v=0$, because $N_{m,0}=\delta_m$ (Exercise 2.5(vi) *Remark*). Since the integral above equals

$$e^{izm}\int_{-\infty}^{\infty} e^{i(z\sqrt{v})y}\frac{1}{\sqrt{2\pi}}e^{-y^2/2}\,dy$$

by the transformation $x = m + \sqrt{v}\,y$, it suffices to check that

$$\frac{1}{\sqrt{2\pi}}\int_{-\infty}^{\infty} e^{izy}e^{-y^2/2}\,dy = e^{-z^2/2}.$$

The left-hand side equals

$$\frac{1}{\sqrt{2\pi}}\int_{-\infty}^{\infty} e^{-z^2/2}e^{-(y+iz)^2/2}\,dy = \frac{1}{\sqrt{2\pi}}e^{-z^2/2}\lim_{a\to\infty}\int_{-a+iz}^{a+iz} e^{-y^2/2}\,dy$$

$$= \frac{1}{\sqrt{2\pi}}e^{-z^2/2}\lim_{a\to\infty}\int_{-a}^{a} e^{-y^2/2}\,dy = e^{-z^2/2}.$$

(Observe the integral along the rectangular route $-a \to a \to a + iz \to -a + iz \to -a$.)

If μ is a distribution, then

$$\check{\mu}(E) = \mu(-E), \qquad -E = \{-x \,|\, x \in E\}$$

is also a distribution, called the *reflection* of μ. It is easy to see that

$$\int_{-\infty}^{\infty} f(x)\check{\mu}(dx) = \int_{-\infty}^{\infty} f(-x)\mu(dx)$$

for every bounded Borel (complex-valued) function f. Putting $f(x) = e^{izx}$, we obtain

$$\mathscr{F}\check{\mu}(z) = \mathscr{F}\mu(-z) = \overline{\mathscr{F}\mu(z)}.$$

More generally, if $L_{ab}(x) = ax + b$, then the image measure μL_{ab}^{-1} is a distribution, for which

$$\int_{\mathbb{R}^1} f(x)(\mu L_{ab}^{-1})(dx) = \int_{\mathbb{R}^1} f(ax+b)\mu(dx)$$

holds. Hence we obtain

$$\mathscr{F}(\mu L_{ab}^{-1})(z) = e^{ibz}(\mathscr{F}\mu)(az).$$

If $\mu = \mu_1 * \mu_2$, we obtain

$$\int_{\mathbb{R}^1} f(x)\mu(dx) = \iint_{\mathbb{R}^2} f(x+y)\mu_1(dx)\mu_2(dy).$$

Putting $f(x) = e^{izx}$ in this formula, we have

$$\mathscr{F}(\mu_1 * \mu_2) = \mathscr{F}\mu_1\mathscr{F}\mu_2$$

For convex combination of distributions we have

$$\mathscr{F}(c_1\mu_1 + c_2\mu_2) = c_1\mathscr{F}\mu_1 + c_2\mathscr{F}\mu_2 \qquad (c_1, c_2 \geq 0, c_1 + c_2 = 1).$$

If $\mu_n \to \mu$, then

$$\int_{\mathbb{R}^1} f(x)\mu_n(dx) \to \int_{\mathbb{R}^1} f(x)\mu(dx)$$

for every bounded continuous (complex-valued) function f. Putting $f(x) = e^{izx}$, we obtain $\mu_n \to \mu \Rightarrow \mathscr{F}\mu_n(z) \to \mathscr{F}\mu(z)$. The following theorem shows that the correspondence $\mu \to \mathscr{F}\mu$ is 1-to-1.

Theorem 2.6.1. If $\varphi = \mathscr{F}\mu$, then

$$\mu(a, b) + \tfrac{1}{2}[\mu\{a\} + \mu\{b\}] = \lim_{c \to \infty} \frac{1}{2\pi} \int_{-c}^{c} \frac{e^{-ibz} - e^{-iaz}}{-iz} \varphi(x)\,dz.$$

Proof: Let $F(c)$ denote the integral on the right-hand side. Then

$$F(c) = \int_{-c}^{c} \int_a^b e^{-ixz}\,dx \int_{-\infty}^{\infty} e^{iyz}\mu(dy)\,dz$$

$$= \int_{-\infty}^{\infty} \mu(dy) \int_a^b dx \int_{-c}^{c} e^{iz(y-x)}\,dz$$

$$= 2\int_{-\infty}^{\infty} \mu(dy) \int_a^b \frac{\sin c(x-y)}{x-y}\,dx$$

$$= 2\int_{-\infty}^{\infty} \mu(dy) \int_{c(a-y)}^{c(b-y)} \frac{\sin u}{u}\,du \qquad \left(x = y + \frac{u}{c}\right)$$

$$= 2\int_{-\infty}^{\infty} [G(c(b-y)) - G(c(a-y))]\mu(dy),$$

where

$$G(x) = \int_0^x \frac{\sin u}{u} \, du.$$

$G(x)$ is a continuous function of x and has finite limits at $\pm \infty$:

$$\lim_{x \to \infty} G(x) = \pi/2, \qquad \lim_{x \to -\infty} G(x) = -\pi/2$$

Hence $G(x)$ is bounded in $-\infty < x < \infty$, so we can exchange the order of integration and limit (as $c \to \infty$) to obtain

$$\lim_{c \to \infty} F(c) = 2 \int_{-\infty}^{\infty} f_{a,b}(x) \mu(dy),$$

where

$$f_{a,b}(x) = \begin{cases} \pi/2 & x = a, b \\ \pi & a < x < b \\ 0 & \text{elsewhere.} \end{cases}$$

Thus we obtain

$$\lim_{c \to \infty} F(c) = 2\pi\mu(a, b) + \pi\mu\{a\} + \pi\mu\{b\}. \qquad \blacksquare$$

This theorem guarantees that $\mu(a, b)$ is determined by $\varphi = F\mu$ for $a, b \in C_\mu$, so that μ is determined by φ. Therefore, we may define a distribution by giving its characteristic function. For example, the Gauss distribution $N_{m,v}$ may be defined to be a distribution with characteristic function $\exp\{imz - vz^2/2\}$. Important distributions and their characteristic functions are shown in Table 2.1.

Since $\mu_1 * \mu_2 = \mu \Leftrightarrow \mathscr{F}\mu_1 \mathscr{F}\mu_2 = \mathscr{F}\mu$, the following relations follow immediately from Table 2.1:

$$b_{n,p} * b_{m,p} = b_{n+m,p},$$

$$p_\lambda * p_\mu = p_{\lambda+\mu},$$

$$\delta_{m_1} * \delta_{m_2} = \delta_{m_1+m_2},$$

$$U_{-c/2,c/2} * U_{-c/2,c/2} = T_{0,c},$$

$$N_{m_1,v_1} * N_{m_2,v_2} = N_{m_1+m_2,v_1+v_2},$$

$$C_{m_1,c_1} * C_{m_2,c_2} = C_{m_1+m_2,c_1+c_2},$$

Table 2.1

Distribution	Characteristic function	
$\check{\mu}$	$\overline{\varphi}$	$(\varphi = \mathscr{F}\mu)$
$\mu * \check{\mu}$	$|\varphi|^2$	$('')$
$\mu L_{ab}^{-1} \quad (a \neq b)$	$e^{ibz}\varphi(az)$	$('')$
$c_1\mu_1 + c_2\mu_2 \quad (c_1, c_2 \geqq 0, c_1 + c_2 = 1)$	$c_1\varphi_1 + c_2\varphi_2$	$(\varphi_i = \mathscr{F}\mu_i)$
$\mu_1 * \mu_2$	$\varphi_1\varphi_2$	$('')$
$\begin{pmatrix} a_n \\ p_n \end{pmatrix}_{n=1,2,\ldots}$	$\sum_n p_n e^{ia_n z}$	
$\mu(dx) = f(x)\,dx$	$\int_{-\infty}^{\infty} e^{ixz}f(x)\,dx$	
$\mu(dx) = f(ax + b)|a|\,dx \quad (a \neq 0)$	$e^{-izb/a}\varphi(z/a)$	$(\varphi = \mathscr{F}(f \cdot dx))$
$b_{n,p} \quad$ (binomial)	$(pe^{iz} + (1-p))^n$	
$p_\lambda \quad$ (Poisson)	$\exp\{\lambda(e^{iz} - 1)\}$	
$\delta_m \quad$ (delta)	e^{imz}	
$U_{a,b} \quad$ (uniform)	$(e^{ibz} - e^{iaz})/iz(b-a)$	
	$= e^{i(a+b)z/2}\dfrac{\sin(b-a)z/2}{(b-a)z/2}$	
$T_{a,c} \quad$ (triangular)	$e^{iaz}\left(\dfrac{\sin zc/2}{zc/2}\right)^2$	
$N_{m,v} \quad$ (Gauss)	$\exp\{imz - \tfrac{1}{2}vz^2\}$	
$C_{m,c} \quad$ (Cauchy)	$\exp\{imz - c|z|\}$	
$E_\lambda \quad$ (exponential)	$(1 - iz/\lambda)^{-1}$	

2.6b   *Elementary properties of characteristic functions*

Let φ be the characteristic function of a distribution μ, namely,

$$\varphi(z) = \int_{-\infty}^{\infty} e^{izx}\mu(dx).$$

Since $|e^{izx}| \leqq 1$, it holds obviously that

$$|\varphi(z)| \leqq 1 = \varphi(0), \qquad \varphi(-z) = \overline{\varphi(z)}.$$

Also we can use the bounded convergence theorem to check the continuity of φ. Furthermore, φ is uniformly continuous by the following:

Theorem 2.6.2.

$$|\varphi(z + h) - \varphi(z)| \leq \sqrt{2|\varphi(0) - \varphi(h)|}\,.$$

Proof: Let $g(z)$ be any integrable function on $\mathbb{R}^1$. The integrals below are taken over the whole real line $\mathbb{R}^1$ unless stated otherwise:

$$\int\int e^{izx}g(z)\,dz\,\mu_n(dx) = \int\int e^{izx}\mu_n(dx)g(z)\,dz \qquad \text{(Fubini's theorem)}$$

$$= \int \varphi_n(z)g(z)\,dz$$

$$\underset{(n\to\infty)}{\rightarrow} \int \varphi(z)g(z)\,dz$$

$$\text{(bounded convergence theorem)}$$

$$= \int\int e^{izx}\mu(dx)g(z)\,dz$$

$$= \int\int e^{izx}g(z)\,dz\,\mu(dx) \qquad \text{(Fubini's theorem)};$$

$$\int h(x)\mu_n(dx) \rightarrow \int h(x)\mu(dx), \qquad \text{where } h(x) = \int e^{izx}g(z)\,dz.$$

Let $f(x)$ be any continuous function with compact support. If we can prove that $f(x)$ is approximated uniformly by the Fourier transforms of integrable functions, then we have

$$\int f(x)\mu_n(dx) \rightarrow \int f(x)\mu(dx),$$

which completes the proof of our theorem by Theorem 2.5.2[(ii) $\Rightarrow$ (i)].
 Define $g_n(z)$ by

$$g_n(z) = \frac{1}{2\pi} e^{-z^2/2n} \int e^{-izy}f(y)\,dy.$$

Then $g_n(z)$ is integrable on $\mathbb{R}^1$, because

$$|g_n(z)| \leq \frac{1}{2\pi} e^{-z^2/2n} \int |f(x)|\,dx.$$

We will prove that

$$h_n(x) = \int e^{izx}g_n(z)\,dz \underset{(n\to\infty)}{\rightarrow} f(x) \qquad \text{(uniform convergence)}.$$

Proof: Using the Schwarz inequality, we obtain

$$|\varphi(z+h)-\varphi(z)|^2 \leq \int_{-\infty}^{\infty} |e^{i(z+h)x}-e^{izx}|^2 \mu(dx)$$

$$= \int_{-\infty}^{\infty} 2(1-\cos hx)\mu(dx)$$

$$= 2\operatorname{Re}(\varphi(0)-\varphi(h)) \leq 2|\varphi(0)-\varphi(h)|. \quad \blacksquare$$

Every characteristic function φ is *positive-definite*, namely,

$$\sum_{j,k=1}^{n} \xi_j \bar{\xi}_k \varphi(z_j - z_k) = \int_{-\infty}^{\infty} \left| \sum_{j=1}^{n} \xi_j e^{iz_j x} \right|^2 \mu(dx) \geq 0,$$

$$n=1,2,\ldots, \xi_1, \xi_2,\ldots, \xi_n \in \mathbf{C}^1, z_1, z_2,\ldots, z_n \in \mathbf{R}^1.$$

Since $\varphi(-z) = \overline{\varphi(z)}$, the matrix $(\varphi(z_j - z_k))_{j,k=1}^n$ is Hermitian. The inequality above shows that this matrix is always positive-definite.

2.6c *Convergence of distributions and characteristic functions*

Let φ_n be the characteristic function of μ_n for $n=1,2,\ldots$. We will investigate the relation between convergence of $\{\mu_n\}$ and that of $\{\varphi_n\}$. Suppose that $\mu_n \to \mu$. Then we have

$$\int_{\mathbf{R}^1} f(x)\mu_n(dx) \to \int_{\mathbf{R}^1} f(x)\mu(dx)$$

for every bounded continuous function $f(x)$ by the definition of convergence of distributions. Putting $f(x) = e^{izx}$, we obtain

$$\varphi_n(z) \to \varphi(z) \equiv \mathscr{F}\mu(z)$$

for each point $z \in \mathbf{R}^1$. In fact we can prove that this convergence is uniform in z (Theorem 2.6.3). To do this we will start with the following lemma, which will frequently be of use.

Lemma 2.6 (Generalization of integration by parts). For every distribution μ and every continuously differentiable function f it holds that

$$\int_{(a,b]} f(x)\mu(dx) = f(b)F(b) - f(a)F(a) - \int_{(a,b]} f'(x)F(x)\,dx,$$

where F is the distribution function of μ.

Proof: Let $g(x, y)$ be any bounded Borel function on $\mathbf{R}^2$ and let μ and ν be any bounded regular measures on $\mathbf{R}^1$. Noticing the obvious equality

$$1_{(a,x]}(y) = 1_{[y,b]}(x) \qquad (a < x, y \leq b)$$

Observe that

$$h_n(x) = \int e^{izx} \frac{1}{2\pi} e^{-z^2/2n} \int e^{-izy} f(y) \, dy \, dz$$

$$= \frac{1}{2\pi} \int\int e^{iz(x-y)} e^{-z^2/2n} \, dz \, f(y) \, dy \qquad \text{(Fubini's theorem)}$$

$$= \sqrt{\frac{n}{2\pi}} \int\int e^{iz(x-y)} \frac{1}{\sqrt{2\pi n}} e^{-z^2/2n} \, dz \, f(y) \, dy$$

$$= \sqrt{\frac{n}{2\pi}} \int e^{-n(x-y)^2/2} f(y) \, dy \qquad \left(\mathscr{F} N_{0,n}(z) = e^{-nz^2/2} \right)$$

$$= \frac{1}{\sqrt{2\pi}} \int e^{-t^2/2} f\left(x + \frac{t}{\sqrt{n}} \right) dt,$$

$$f(x) = \frac{1}{\sqrt{2\pi}} \int e^{-t^2/2} f(x) \, dt,$$

$$|h_n(x) - f(x)| \leq \frac{1}{\sqrt{2\pi}} \int e^{-t^2/2} \left| f\left(x + \frac{t}{\sqrt{n}} \right) - f(x) \right| dt$$

and

$$\sup_x |h_n(x) - f(x)| \leq \frac{1}{\sqrt{2\pi}} \int e^{-t^2/2} \sup_x \left| f\left(x + \frac{t}{\sqrt{n}} \right) - f(x) \right| dt.$$

Since $f(x)$ is bounded and uniformly continuous on $\mathbb{R}^1$, we can use the dominated convergence theorem to obtain

$$\sup_x |h_n(x) - f(x)| \to 0 \qquad \blacksquare$$

Let φ_n be the characteristic function of a distribution μ_n for $n = 1, 2, \ldots,$ and suppose that $\varphi_n(z)$ converges to a function $\varphi(z)$ for every point z. If $\varphi(z)$ is the characteristic function of a distribution μ, then μ_n converges to μ by Glivenko's theorem. But this is not the case in general, as we can see in the following example:

Let $\mu_n = N_{0,n}$. Then

$$\varphi_n(z) = \mathscr{F} N_{0,n}(z) = e^{-nz^2/2} \to \varphi(z) = \begin{cases} 1, & z = 0, \\ 0, & z \neq 0. \end{cases}$$

$\varphi(z)$ is not a characteristic function because of its discontinuity at $z = 0$.

The following theorem gives a condition for the limit function φ to be a characteristic function.

Theorem 2.6.5 (Lévy's convergence theorem). If $\varphi_n(z) = \mathscr{F}\mu_n(z)$ converges to $\varphi(z)$ at every point z and if this convergence is uniform in a neighborhood $|z| < a$, where a may be any small number, then the limit function φ is the characteristic function of a distribution μ. Therefore $\mu_n \to \mu$ by Glivenko's theorem.

Proof: Since φ_n is continuous, $\varphi(z)$ is continuous at $z = 0$ by the assumption of uniform convergence. Since

$$\varphi(0) = \lim_{n \to \infty} \varphi_n(0) = 1,$$

we can find, for every $\varepsilon > 0$, $\delta = \delta(\varepsilon) \in (0, a)$ such that $|\varphi(z) - 1| < \varepsilon$ whenever $|z| < \delta$. Since $\varphi_N(z)$ converges to $\varphi(z)$ uniformly in $|z| < \delta$, we can find $N = N(\varepsilon)$ (independent of z) such that $|\varphi_n(z) - 1| < \varepsilon$ whenever $n > N$ and $|z| < \delta$. Hence, whenever $n > N$, we have

$$1 - \varepsilon < \frac{1}{2\delta} \int_{-\delta}^{\delta} \mathrm{Re}(\varphi_n(z))\, dz = \frac{1}{2\delta} \int_{-\delta}^{\delta} \int_{-\infty}^{\infty} (\cos zx)\mu_n(dx)\, dz$$

$$= \int_{-\infty}^{\infty} \frac{\sin \delta x}{\delta x} \mu_n(dx), \quad \text{(Fubini's theorem)}.$$

Since the absolute value of the integrand of the last integral is bounded by 1 on $[-b, b]$ and by $1/\delta b$ elsewhere, where b is any positive number, this integral is bounded by $\mu_n[-b, b] + 1/\delta b$. Hence we have

$$\mu_n[-b, b] \geq 1 - \varepsilon - 1/\delta b, \qquad n > N,$$

namely,

$$\inf_{n > N} \mu_n[-b, b] \geq 1 - \varepsilon - 1/\delta b,$$

which implies that

$$\lim_{b \to \infty} \inf_{n > N} \mu_n[-b, b] \geq 1 - \varepsilon.$$

Since $\lim_{b \to \infty} \mu_n[-b, b] = 1$ for each n, it holds that

$$\lim_{b \to \infty} \inf_{n \leq N} \mu_n[-b, b] = 1 \geq 1 - \varepsilon.$$

Summarizing the results above, we obtain

$$\lim_{b \to \infty} \inf_n \mu_n[-b, b] \geq 1 - \varepsilon \to 1 \qquad (\varepsilon \to 0).$$

Therefore $\{\mu_n\}$ has a convergent subsequence $\{\nu_n\}$ by Theorem 2.5.3. Denoting the limit distribution by μ, we see that $\mathscr{F}\nu_n(z) \to \mathscr{F}\mu(z)$ holds for every z. Since $\mathscr{F}\nu_n$ is a subsequence of $\varphi_n = \mathscr{F}\mu_n$ and since $\varphi_n(z) \to \varphi(z)$, we have $\varphi = \mathscr{F}\mu$. ∎

2.6.d   Bochner's theorem

A characteristic function is the Fourier transform of a distribution μ by definition. It is often the case that we have to decide that a given function is a characteristic function by the properties of the function itself without referring to the corresponding distribution. The following theorem is useful for this purpose.

Theorem 2.6.6 (Bochner's theorem). In order for $\varphi(z)$ to be a characteristic function it is necessary and sufficient that the following three conditions hold:

(B.1) φ is positive-definite.

(B.2) $\varphi(z)$ is continuous at $z = 0$.

(B.3) $\varphi(0) = 1$.

Proof: The necessity is obvious. We will prove the sufficiency. Suppose that φ satisfies the three conditions above. Condition (B.1) implies that

$$\sum_{j,k=1}^{n} \xi_j \bar{\xi}_k \varphi(z_j - z_k) \geqq 0, \qquad n = 1, 2, \ldots,$$

$$\xi_1, \xi_2, \ldots, \xi_n \in \mathbb{C}^1, \qquad z_1, z_2, \ldots, z_n \in \mathbb{R}^1.$$

Setting $n = 2$, $\xi_1 = \xi_2 = 1$, $z_1 = z$, $z_2 = 0$ we have $2\varphi(0) + \varphi(z) + \varphi(-z) \geqq 0$, which implies that $\varphi(z) + \varphi(-z)$ is real. Setting $n = 2$, $\xi_1 = i$, $\xi_2 = 1$, $z_1 = z$, $z_2 = 0$ we have $2\varphi(0) = i\varphi(z) - i\varphi(-z) \geqq 0$, which implies that $\varphi(z) - \varphi(-z)$ is purely imaginary. Therefore, we obtain $\varphi(-z) = \overline{\varphi(z)}$. Keeping this in mind and setting $n = 2$, $\xi_1 = \xi$, $\xi_2 = \eta$, $z_1 = z$, $z_2 = 0$ we have $\xi\bar{\xi} + \xi\bar{\eta}\varphi(z) + \bar{\xi}\eta\overline{\varphi(z)} + \eta\bar{\eta} \geqq 0$. This implies that

$$\begin{bmatrix} 1 & \overline{\varphi(z)} \\ \varphi(z) & 1 \end{bmatrix}$$

is a positive-definite Hermitian matrix, so

$$\begin{vmatrix} 1 & \overline{\varphi(z)} \\ \varphi(z) & 1 \end{vmatrix} \geqq 0,$$

namely, $|\varphi(z)| \leqq 1$. Hence we see that φ is a bounded function. Setting $n = 3$, $\xi_1 = \xi$, $\xi_2 = -\xi$, $\xi_3 = \eta$, $z_1 = 0$, $z_2 = h$, $z_3 = z + h$ and using assumption (B.3), we have

$$\xi\bar{\xi}\left(2 - \varphi(h) - \overline{\varphi(h)}\right) + \xi\bar{\eta}\left(\overline{\varphi(z+h)} - \overline{\varphi(z)}\right)$$

$$+ \bar{\xi}\eta\left(\varphi(z+h) - \varphi(z)\right) + \eta\bar{\eta} \geqq 0.$$

By the same argument as above we can deduce from this the following inequality:

$$|\varphi(z+h)-\varphi(z)|^2 \leq 2-\varphi(h)-\overline{\varphi(h)} = 2\,\mathrm{Re}(1-\varphi(h)) \leq 2|1-\varphi(h)|,$$

which implies the uniform continuity because of assumption (B.2). Thus we have seen that $\varphi(z)$ is bounded and uniformly continuous.

If g is integrable, bounded, and uniformly continuous, then

$$\iint \varphi(t-s)g(t)\overline{g(s)}\,dt\,ds$$

$$= \lim_{n\to\infty} \sum_{j,k=-n^2}^{n^2} \varphi\!\left(\frac{j}{n}-\frac{k}{n}\right) g\!\left(\frac{j}{n}\right)\overline{g\!\left(\frac{k}{n}\right)}\!\left(\frac{1}{n}\right)^2 \geq 0,$$

namely,

$$\int \varphi(t)\int g(t+s)\overline{g(s)}\,ds\,dt \geq 0.$$

Setting

$$g(t) = N_{0,\,n/4}(t)e^{-ixt}$$

and noting

$$N_{0,\,v}(t) = N_{0,\,v}(-t), \qquad N_{0,\,\alpha}*N_{0,\,\beta} = N_{0,\,\alpha+\beta},$$

we obtain

$$\int g(t+s)\overline{g(s)}\,ds = N_{0,\,n/2}(t)e^{-ixt} = \sqrt{\frac{1}{\pi n}}\,e^{-t^2/n}e^{-ixt},$$

and so

$$(2\pi)^{-1}\int_{-\infty}^{\infty} \varphi(t)e^{-t^2/n}e^{-ixt}\,dt \geq 0.$$

Denoting the left-hand side by $f_n(x)$, we will prove that

$$\mu_n(dx) = f_n(x)\,dx$$

defines a distribution. It suffices to show that $\mu_n(\mathbb{R}^1)=1$ because μ_n is a measure on $\mathscr{B}^1$ obviously. Observe that

$$\mu_n(-a,\,a) = \int_{-a}^{a} f_n(x)\,dx$$

$$= \frac{1}{2\pi}\int_{-a}^{a}\int_{-\infty}^{\infty} \varphi(t)e^{-t^2/n}e^{-ixt}\,dt\,dx,$$

$$= \frac{1}{2\pi}\int_{-\infty}^{\infty} \varphi(t)e^{-t^2/n}\frac{2\sin at}{t}\,dt \qquad \text{(Fubini's theorem)}.$$

Since $\mu_n(-a, a)\uparrow\mu_n(\mathbb{R}^1)$ as $a\to\infty$, we have

$$\mu_n(\mathbb{R}^1) = \lim_{b\to\infty}\frac{1}{b}\int_0^b\mu_n(-a, a)\,da$$

$$= \lim_{b\to\infty}\frac{1}{b}\int_0^b\frac{1}{2\pi}\int_{-\infty}^\infty\varphi(t)e^{-t^2/n}\frac{2\sin at}{t}\,dt\,da$$

$$= \lim_{b\to\infty}\frac{1}{2\pi}\int_{-\infty}^\infty\varphi(t)e^{-t^2/n}\frac{2(1-\cos tb)}{t^2 b}\,dt$$

$$= \lim_{b\to\infty}\frac{1}{2\pi}\int_{-\infty}^\infty\varphi\left(\frac{t}{b}\right)e^{-t^2/nb^2}\frac{2(1-\cos t)}{t^2}\,dt.$$

Noting that

$$\frac{1-\cos t}{t^2}\geq 0,\qquad\int_{-\infty}^\infty\frac{1-\cos t}{t^2}\,dt=\pi,\text{ and }\varphi(0)=1,$$

we can use the dominated convergence theorem to conclude that this limit equals 1. Thus we have seen that μ_n is a distribution.

Next we will prove that the characteristic function $\varphi_n(z)=\mathscr{F}\mu_n(z)$ equals $\varphi(z)e^{-z^2/n}$. Observe that

$$\varphi_n(z) = \lim_{a\to\infty}\int_{-a}^a e^{izx}\frac{1}{2\pi}\int_{-\infty}^\infty\varphi(t)e^{-t^2/n}e^{-itx}\,dt\,dx$$

$$= \lim_{a\to\infty}\frac{1}{2\pi}\int_{-\infty}^\infty\varphi(t)e^{-t^2/n}\frac{2\sin a(t-z)}{t-z}\,dt$$

$$= \lim_{b\to\infty}\frac{1}{b}\int_0^b da\frac{1}{2\pi}\int_{-\infty}^\infty\varphi(t)e^{-t^2/n}\frac{2\sin a(t-z)}{t-z}\,dt$$

$$= \lim_{b\to\infty}\frac{1}{2\pi}\int_{-\infty}^\infty\varphi(t)e^{-t^2/n}\frac{2(1-\cos b(t-z))}{b(t-z)^2}\,dt$$

$$= \lim_{b\to\infty}\frac{1}{2\pi}\int_{-\infty}^\infty\varphi\left(z+\frac{s}{b}\right)e^{-(z+s/b)^2/n}\frac{2(1-\cos s)}{s^2}\,ds$$

$$= \varphi(z)e^{-z^2/n}.$$

Therefore $\varphi_n(z)\to\varphi(z)$ $(n\to\infty)$ for every z and the convergence is obviously uniform in $|z|\leq 1$. Hence Lévy's convergence theorem ensures that φ is a characteristic function. ∎

Exercise 2.6.

(i) Let $f(x)$ be an integrable complex-valued function on $\mathbb{R}^1$. We define the *Fourier transform* $\varphi = \mathscr{F}f$ by

$$\varphi(t) = \int_{-\infty}^{\infty} e^{ixt} f(x)\, dx.$$

Show that if f is continuous and if φ is integrable, then

$$f(x) = \frac{1}{2\pi} \int_{-\infty}^{\infty} e^{-ixt} \varphi(t)\, dt.$$

[*Hint:*

$$\int_{-\infty}^{\infty} e^{-ixt} \varphi(t)\, dt = \lim_{a \to \infty} \int_{-a}^{a} e^{-ixt} \varphi(t)\, dt$$

$$= \lim_{b \to \infty} \frac{1}{b} \int_{0}^{b} da \int_{-a}^{a} e^{-ixt} \varphi(t)\, dt.$$

$$= \lim_{b \to \infty} \frac{1}{b} \int_{0}^{b} da \int_{-\infty}^{\infty} \frac{2\sin((y-x)a)}{(y-x)} f(y)\, dy.]$$

(ii) Prove that $f(x) = (2\pi)^{-1}((\sin(x/2))/(x/2))^2$ is the density of a distribution and that the characteristic function of the distribution is

$$T_{0,1}(z) = \begin{cases} 1 - |z|, & |z| \leq 1 \\ 0, & |z| > 1 \end{cases} = (1 - |z|)^+.$$

[*Hint:* Note that $\mathscr{F}T_{0,1} = ((\sin(x/2))/(x/2))^2$ and use (i).]

(iii) Suppose that we are given a family of functions $F_a(\xi)$, $\xi \in \Xi$, indexed by a point a of a probability space (A, ν) and that $F_a(\xi)$ is integrable in a for every ξ. Then the function

$$F(\xi) = \int_A F_a(\xi)\nu(da), \qquad \xi \in \Xi$$

is called the *integral convex combination* of $\{F_\alpha,\ \alpha \in A\}$ with weight measure ν. Prove that every integral convex combination of a family of distributions viewed as functions on $\mathbb{B}^1$ is also a distribution and that the same fact holds for characteristic functions.

Use this to prove the following fact (*Pólya's theorem*): If $\varphi(z)$ is a nonnegative even function convex in $(0, \infty)$ and continuous at $z = 0$ and if $\varphi(0) = 1$, then $\varphi(z)$ is a characteristic function.

[*Hint*: Since $\psi(z/a)$ $(a > 0)$ is a characteristic function with $\psi(z)$,

$$\varphi_a(z) = \left(1 - \left|\frac{z}{a}\right|\right)^+, \qquad 0 < a < \infty$$

is a characteristic function by (i). Also $\varphi_\infty(z) \equiv 1$ is the characteristic function of δ. Hence it suffices to check that $\varphi(z)$ is an integral convex combination of $\varphi_a(z)$, $0 < a \leq \infty$. The weight measure ν is obtained from

$$\varphi(z) = \int_{(0,\infty]} \varphi_a(z)\nu(da)$$

as follows. For $z > 0$ we have

$$\varphi(z) = \int_{[z,\infty)} \left(1 - \frac{z}{a}\right)\nu(da) + \nu\{\infty\}$$

$$= \int_{[z,\infty)} \int_{[z,a)} d\xi \frac{1}{a} \nu(da) + \nu\{\infty\}$$

$$= \int_{[z,\infty)} \int_{(\xi,\infty)} \frac{1}{a} \nu(da)\, d\xi + \nu\{\infty\}$$

and so

$$\nu\{\infty\} = \lim_{z \to \infty} \varphi(z).$$

Also, we have

$$\varphi^+(z)(\text{right derivative}) = -\int_{(z,\infty)} \frac{1}{a}\nu(da),$$

$$d\varphi^+(z) = \frac{1}{z}\nu(dz),$$

$$\nu(dz) = zd\varphi^+(z) \geq 0 \qquad (\text{by convexity of } \varphi).$$

It remains only to prove that

$$\int_{(0,\infty)} ad\varphi^+(a) + \nu\{\infty\} = 1,$$

$$\int_{(0,\infty)} \varphi_a(z)ad\varphi^+(a) + \nu\{\infty\}\varphi_\infty(z) = \varphi(z).$$

The first formula is obtained from the second one by putting $z = 0$. To prove the second one, use

$$\varphi^+(\infty) = 0, \qquad \varphi_\infty(z) \equiv 1$$

and observe that

$$\int_{(0,\infty)} \varphi_a(z) a \, d\varphi^+(a) = \int_{[z,\infty)} (a-z) \, d\varphi^+(a)$$

$$= \int_{[z,\infty)} \int_{[z,a)} d\xi \, d\varphi^+(a)$$

$$= \varphi(z) - \nu\{\infty\} \qquad \text{(Fubini's theorem).]}$$

(v) Use the residue theorem to compute the characteristic function of Cauchy distribution.

(vi) Use Bochner's theorem to prove that if a sequence of characteristic functions $\{\varphi_n(z)\}$ converges to $\varphi(z)$ at every point z and if $\varphi(z)$ is continuous at $z = 0$, then $\varphi(z)$ is a characteristic function. (This fact implies Lévy's convergence theorem.)

(vii) Prove that if $|M|_n(\mu) < \infty$, then $\varphi(z) = \mathscr{F}\mu(z)$ is n times continuously differentiable and

$$\varphi^{(k)}(z) = \int_{R^1} (ix)^k e^{izx} \mu(dx) \qquad (k = 0,1,2,\ldots,n; \ \varphi^{(0)} \equiv \varphi).$$

[*Hint*: $|M|_k(\mu) < \infty$, $k = 1,2,\ldots,n$, by the assumption. Hence the integral above is well-defined and is continuous in z by the bounded convergence theorem. The equality is obvious for $k = 0$. If it holds for $k(< n)$, then it holds for $k+1$ as we can see below:

$$\frac{1}{h}\left(\varphi^{(k)}(z+h) - \varphi^{(k)}(z)\right) = \int_{-\infty}^{\infty} \frac{e^{i(z+h)x} - e^{izx}}{h} (ix)^k \mu(dx)$$

$$= \frac{1}{h} \int_{-\infty}^{\infty} \int_{z}^{z+h} e^{i\xi x} \, d\xi \, (ix)^{k+1} \mu(dx)$$

$$= \frac{1}{h} \int_{z}^{z+h} d\xi \int_{-\infty}^{\infty} e^{i\xi x} (ix)^{k+1} \mu(dx)$$

$$\text{(Fubini's theorem)}$$

$$\rightarrow \int_{-\infty}^{\infty} e^{izx} (ix)^{k+1} \mu(dx) \qquad (h \rightarrow 0).]$$

(viii) Prove that if $\varphi(z) = \mathscr{F}\mu(z)$ is $2n$ times differentiable at $z = 0$, then $|M|_{2n}(\mu) < \infty$ and $\varphi^{(k)}(z)$ $(k = 0,1,2,\ldots,2n)$ are given by the formula of (vii). Setting $z = 0$, we have

$$M_k(\mu) = i^{-k}\varphi^{(k)}(0), \qquad k = 0,1,2,\ldots,2n.$$

[*Hint*: First consider the case where $k = 1$. Observe

$$-\frac{1}{h^2}(\varphi(h) - 2\varphi(0) + \varphi(-h)) = \int_{-\infty}^{\infty} \frac{2(1 - \cos hx)}{h^2} \mu(dx)$$

$$\geq \int_{-A}^{A} \frac{2(1 - \cos hx)}{h^2} \mu(dx)$$

$$= \frac{4}{h^2} \int_{-A}^{A} \left(\sin\frac{hx}{2}\right)^2 \mu(dx)$$

$$\geq \frac{4}{h^2} \int_{-A}^{A} \left(\frac{2}{\pi}\right)^2 \left(\frac{hx}{2}\right)^2 \mu(dx)$$

(whenever $|h| < \pi/A$)

$$= \left(\frac{2}{\pi}\right)^2 \int_{-A}^{A} x^2 \mu(dx),$$

and let $h \downarrow 0$ and then $A \uparrow \infty$ to obtain

$$-\varphi''(0) \geq \left(\frac{2}{\pi}\right)^2 \int_{-A}^{A} x^2 \mu(dx),$$

$$-\varphi''(0) \geq \left(\frac{2}{\pi}\right)^2 \int_{-\infty}^{\infty} x^2 \mu(dx).$$

This implies that $|M|_2(\mu) < \infty$. Hence it follows by (vii) that

$$\varphi^{(2)}(z) = \int_{-\infty}^{\infty} e^{izx}(ix)^2 \mu(dx) \qquad (z \in \mathbb{R}^1).$$

Use induction for the general case.]
(ix) Use (viii) to prove that

$$M_{2p-1}(N_{0,1}) = 0 \qquad \text{and} \qquad M_{2p}(N_{0,1}) = 1 \cdot 3 \cdot 5 \cdots (2p - 1).$$

[*Hint*:

$$\mathcal{F}N_{0,1}(z) = e^{-z^2/2} = \sum_{v=0}^{\infty} \left(-\frac{1}{2}\right)^p \frac{z^{2p}}{p!} .]$$

2.7   The weak topology in the distributions

Let $\mathscr{P}^1$ denote the distributions. We have defined limits of sequences in $\mathscr{P}^1$ in Subsection 2.5.b. In this section we will introduce a Hausdorff topology τ in $\mathscr{P}^1$ such that $\mu_n \to \mu(\tau)$ if and only if $\mu_n \to \mu$ in the sense mentioned above. Let C_b denote the class of all bounded continuous real functions on $\mathbb{R}^1$ and C_K the class of all functions that belong to C_b and have compact support. If we set

$$(f, \mu) = \int_{\mathbb{R}^1} f(x)\mu(dx), \quad f \in C_b, \mu \in \mathscr{P}^1,$$

then (f, μ) is a bilinear form. Defining the following class of sets to be a neighborhood system of $\mu \in \mathscr{P}^1$

$$U_{f_1, f_2, \ldots, f_n, \varepsilon}(\mu) = \left\{ v \in \mathscr{P}^1 \| (f_k, v) - (f_k, \mu) | < \varepsilon, k = 1, 2, \ldots, n \right\}$$

$$(f_1, f_2, \ldots, f_n \in C_b; \varepsilon > 0; n = 1, 2, \ldots),$$

we obtain a Hausdorff topology in $\mathscr{P}^1$, denoted by τ_b. This topology is called the *weak topology* in $\mathscr{P}^1$. It is obvious that $\mu_n \to \mu$ in the sense of Subsection 2.5.b if and only if $\mu_n \to \mu(\tau_b)$.

A new Hausdorff topology τ_K can be defined similarly by replacing C_b by C_K. This topology τ_K appears to be weaker than τ_b. But these two topologies coincide with each other. To prove this, it suffices to check that for any $\varepsilon > 0$ and any $f_1, f_2, \ldots, f_n \in C_b$ we can find $\delta > 0$ and $g_0, g_1, \ldots, g_n \in C_K$ such that

$$U_{g_0, g_1, \ldots, g_n, \delta}(\mu) \subset U_{f_1, f_2, \ldots, f_n, \varepsilon}(\mu).$$

Define $e_A \in C_K$ by

$$e_A(x) = \begin{cases} 1, & |x| \leq A, \\ 0, & |x| \geq A + 1, \\ \text{linear}, & x \in [-A-1, -A] \quad \text{or} \quad x \in [A, A+1]. \end{cases}$$

Then it holds for every $v \in \mathscr{P}^1$ that

$$|(e_A f_k, v) - (f_k, v)| = |((1 - e_A)f_k, v)|$$

$$\leq |(\|f_k\|_\infty (1 - e_A), v)| \quad \left(\|f_k\|_\infty = \sup_x |f_k(x)| \right)$$

$$\leq \|f_k\|_\infty (1 - e_A, v)$$

$$\leq \|f_k\|_\infty (1 - (e_A, \mu)) + \|f_k\|_\infty |(e_A, v) - (e_A, \mu)|$$

$$\leq \|f_k\|_\infty \mu([-A, A]^c) + \|f_k\|_\infty |(e_A, v) - (e_A, \mu)|$$

$$\leq a(\mu([-A, A]^c) + |(e_A, v) - (e_A, \mu)|)$$

$$\left(a = \max_k \|f_k\|_\infty \right).$$

Setting $\nu = \mu$, we have

$$|(e_A f_k, \mu) - (f_k, \mu)| \leq a\mu([-A, A]^c).$$

and so

$$|(f_k, \nu) - (f_k, \mu)| \leq |(e_A f_k, \nu) - (e_A f_k, \mu)| + 2a\mu([-A, A]^c)$$
$$+ a|(e_A, \nu) - (e_A, \mu)|.$$

Taking $A = A(\varepsilon)$ sufficiently large so that

$$2a\mu([-A, A]^c) < \varepsilon/2,$$

we can check that

$$|(f_k, \nu) - (f_k, \mu)| < \varepsilon, \qquad k = 1, 2, \ldots, n,$$

whenever

$$|(e_A, \nu) - (e_A, \mu)| < \varepsilon/4a, \qquad |(e_A f_k, \nu) - (e_A f_k, \mu)| < \varepsilon/4,$$
$$k = 1, 2, \ldots, n.$$

This implies that

$$U_{g_0, g_1, \ldots, g_n, \delta}(\mu) \subset U_{f_1, f_2, \ldots, f_n, \varepsilon}(\mu),$$

where

$$g_0 = e_A, \qquad g_k = e_A f_k \ (k = 1, 2, \ldots, n), \qquad \text{and} \qquad \delta = \varepsilon/(4a + 4).$$

(From now on we denote the weak topology by τ deleting the subscript.)

The topology τ is metrizable. To prove this, we take a sequence $\{g_k\}$ dense in C_K with respect to the supremum norm and define a metric ρ in $\mathscr{P}^1$ by

$$\rho(\mu, \nu) = \sum_{n=1}^{\infty} 2^{-n}(|(g_n, \mu) - (g_n, \nu)| \wedge 1), \qquad a \wedge b = \min(a, b).$$

Then the ρ topology in $\mathscr{P}^1$ coincides with the weak topology τ.

Before proving this we will prove the existence of a sequence $\{g_k\}$ satisfying the above-mentioned condition. Take an arbitrary set of rational numbers

$$-n = a_0 < a_1 < \cdots < a_m = n$$

and

$$0 = b_0, b_1, \ldots, b_m = 0$$

and define a function $g(x)$ that takes b_i at each a_i, is linear on each interval $[a_{i-1}, a_i]$, and vanishes elsewhere. The set of all such functions is countable and can be denoted by $\{g_1, g_2, \ldots\}$.

The weak topology is stronger then the ρ topology, for if $\sum_{n=N+1}^{\infty} 2^{-n} < \varepsilon$ and $\nu \in U_{g_1, g_2, \ldots, g_{N\varepsilon}}(\mu)$, then $\rho(\mu, \nu) < 2\varepsilon$. For any given functions $f_1, f_2, \ldots, f_n \in C_K$ and any $\varepsilon \in (0, \frac{1}{2})$ we take N sufficiently large so that every f_k is ε-approximated by $g_p(p = p(k) \le N)$ with respect to the supremum norm. Then we have, for every $\lambda \in \mathscr{P}^1$,

$$\left| (f_k, \lambda) - (g_{p(k)}, \lambda) \right| \le \| f_k - g_{p(k)} \|_\infty < \varepsilon$$

and

$$|(g_{p(k)}, \lambda)| \le |(f_k, \lambda)| + \varepsilon < \| f_k \|_\infty + \tfrac{1}{2},$$

so

$$\begin{aligned} |(f_k, \nu) - (f_k, \mu)| &\le 2\varepsilon + |(g_{p(k)}, \nu) - (g_{p(k)}, \mu)| \\ &= 2\varepsilon + |(g_{p(k)}, \nu) - (g_{p(k)}, \mu)| \wedge (2a + 1) \\ &\le 2\varepsilon + (2a + 1)\big(|(g_{p(k)}, \nu) - (g_{p(k)}, \mu)| \wedge 1\big) \\ &\le 2\varepsilon + (2a + 1)2^N \rho(\mu, \nu), \end{aligned}$$

where $a = \max_k \| f_k \|_\infty$, $k = 1, 2, \ldots, n$. This implies that $\nu \in U_{f_1, f_2, \ldots, f_n, 3\varepsilon}(\mu)$ if $\rho(\mu, \nu)$ is sufficiently small, proving that the ρ topology is stronger than the weak topology.

Thus the weak topology τ is determined by the metric ρ and therefore satisfies the first countability axiom. Moreover it satisfies the second countability axiom. To prove this it suffices to note that the distributions of the form

$$\begin{pmatrix} a_1, a_2, \ldots, a_n \\ p_1, p_2, \ldots, p_n \end{pmatrix}, \qquad (n = 1, 2, \ldots; \ a_i, p_i: \text{rational})$$

are dense in $\mathscr{P}^1$.

The Hausdorff topological space $\mathscr{P}^1$ with the weak topology τ is a *Polish space*. In other words τ is given by a metric d in such a way that the metric space $(\mathscr{P}^1, d)$ is a complete separable metric space. The metric ρ introduced above does not satisfy this condition, because the metric space $(\mathscr{P}^1, \rho)$ is separable but not complete. Hence we have to seek a new metric d_L that determines the weak topology in order to prove that $(\mathscr{P}^1, \tau)$ is Polish. Such a metric was introduced by P. Lévy and is called the *Lévy metric*.

Let μ be a distribution and F its distribution function. Consider the graph of $y = F(x)$ and connect $(a, F(a-))$ with $(a, F(a+))$ by a straight line segment at each discontinuity point a to obtain a continuous curve in the $x-y$ plane (this will be called the graph of F in the discussion below). Let $G(t) = G_\mu(t)$ denote the distance between $(t, 0)$ and the intersection of the straight line $x + y = t$ with the graph of F. (See Figure 2.2.) Then $G(t)$ is an increasing function of t satisfying

$$|G(t) - G(s)| \leq \sqrt{2}\,|t - s|, \qquad \lim_{t \to -\infty} G(t) = 0; \ \lim_{t \to \infty} G(t) = \sqrt{2}.$$

The correspondence $\mu \to G_\mu$ is clearly 1-to-1. Let $\mathcal{G}$ denote the set of all such functions G's. We introduce a metric δ in $\mathcal{G}$ by

$$\delta(G_1, G_2) = \sup_t |G_1(t) - G_2(t)| \equiv \|G_1 - G_2\|_\infty$$

and the Lévy metric d_L in $\mathcal{P}^1$ by

$$d_L(\mu, \nu) = \delta(G_\mu, G_\nu).$$

As the metric space $(\mathcal{G}, \delta)$ is obviously complete and separable, so is the metric space $(\mathcal{P}^1, d_L)$.

To prove that $(\mathcal{P}^1, \tau)$ is Polish, it remains only to check that the d_L topology coincides with τ. Since τ is determined by the metric d observed above, it suffices to show that

$$\mu_n \to \mu(\tau) \Leftrightarrow d_L(\mu_n, \mu) \to 0,$$

namely, that $\|G_n - G\|_\infty \to 0$ if and only if $F_n(a) \to F(a)$ at every continuity point a of F, where (F_n, G_n) and (F, G) correspond to μ_n and μ, respectively.

Suppose that $\|G_n - G\|_\infty \to 0$ and let a be a continuity point of F. Denote by (a_n, b_n) the intersection of the straight line $x + y = a + F_n(a)$

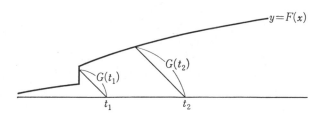

Figure 2.2

with the graph of F. If $a_n \geq a$, then

$$0 \leq F_n(a) - F(a) \leq (F_n(a) - b_n) + (b_n - F(a))$$
$$\leq \|G_n - G\|_\infty + (F(a_n) - F(a)) \to 0.$$

Also, if $a_n < a$, then

$$0 \leq F(a) - F_n(a) \leq (F(a) - b_n) + (b_n - F_n(a))$$
$$\leq (F(a) - F(a_n -)) + \|G_n - G\|_\infty \to 0.$$

Suppose conversely that $F_n(a) \to F(a)$ at every continuity point a of F. Take continuity points of F, $a_0, a_1, \ldots, a_m$, such that

$$F(a_0) < \varepsilon, \qquad F(a_m) > 1 - \varepsilon, \qquad 0 < a_k - a_{k-1} < \varepsilon; \qquad k = 1, 2, \ldots, m.$$

Then

$$|F_n(a_k) - F(a_k)| < \varepsilon, \qquad k = 0, 1, 2, \ldots, m$$

for n sufficiently large. Denote by (ξ_n, η_n) and (ξ, η) the intersections of the straight line $x + y = t$ with the graph of F_n and that of F, respectively. If $\xi < a_0$ or $\xi > a_m$, then the distance between (ξ_n, η_n) and (ξ, η) is less than $2\sqrt{2}\,\varepsilon$. If $a_{k-1} < \xi < a_k$, then the distance is less than $\sqrt{2}\,\varepsilon + 2\sqrt{2}\,\varepsilon$, as we can see from the graphs of F_n and F. Hence we obtain

$$\|G_n - G\|_\infty < \sqrt{2}\,\varepsilon + 2\sqrt{2}\,\varepsilon,$$

Exercise 2.7.
  (i) Compute $d_L(T_{0,c}, \delta)$.
  (ii) Prove that the following three conditions are equivalent to each other, where $\mathscr{M}$ is a subset of $\mathscr{P}^1$.
    (a) The τ-closure of $\mathscr{M}$ is compact.
    (b) $\mathscr{M}$ is totally bounded with respect to the Lévy metric; namely, $\mathscr{M}$ is convered by a finite number of ε-balls with respect to d_L for every $\varepsilon > 0$.
    (c) $\lim_{a \to \infty} \inf_{\mu \in \mathscr{M}} \mu[-a, a] = 1$.

2.8   *d*-Dimensional distributions

A regular probability measure on $\mathbb{R}^d$ is called a *d-dimensional distribution*. The set of all distributions on $\mathbb{R}^d$ is denoted by $\mathscr{P}^d$ similarly to the one-dimensional case. All the properties of one-dimensional distributions discussed in the previous section can be generalized to d-dimensional ones.

Let us remark that the complete direct product of d_i-dimensional distributions μ_i $(i = 1, 2, \ldots, n)$ is a d-dimensional distribution, where $d = \sum_i d_i$. Hence, if $\mu_1, \mu_2, \ldots, \mu_d$ are one-dimensional distributions, then

$$\mu = \overline{\mu_1 \times \mu_2 \times \cdots \times \mu_d} \qquad \text{(complete direct product)}$$

is a d-dimensional distribution. In particular, if

$$\mu_i = \begin{pmatrix} a_{i1}, a_{i2}, \ldots \\ p_{i2}, p_{i2}, \ldots \end{pmatrix}, \qquad i = 1, 2, \ldots, d,$$

then μ is given by

$$\mu = \begin{pmatrix} (a_{1j_1}, a_{2j_2}, \ldots, a_{dj_d}) \\ p_{1j_1} p_{2j_2} \cdots p_{dj_d} \end{pmatrix}_{j_1, j_2, \ldots, j_d}$$

If every μ_i has density f_i then μ has density

$$f(x_1, x_2, \ldots, x_d) = f_1(x_1) f_2(x_2) \ldots f_d(x_d).$$

If every μ_i is singular, then μ is also singular.

The Lebesgue decomposition theorem also holds for d-dimensional distributions. The absolutely continuous distributions coincide with the distributions with density. The Gauss distribution $N_{m,v}$ is the most famous of such distributions and its density f is given by

$$f(x_1, x_2, \ldots, x_d) = \frac{1}{(2\pi)^{d/2}\sqrt{\det V}} \exp\left\{ -\frac{1}{2}{}^t(x - m)V^{-1}(x - m) \right\},$$

where $m \in \mathbb{R}^d$ and V is a strictly positive-definite real matrix and $x - m$ is a column vector and ${}^t(x - m)$ is its transpose ($1 \times d$-matrix). ${}^t(x - m)V^{-1}(x - m)$ may be written in the inner product form:

$$\left(V^{-1}(x - m), (x - m) \right)$$

To see that f is a density function it suffices to prove that the integral of f over $\mathbb{R}^1$ is equal to 1. By the assumption we can write V as follows:

$$V = UDU^{-1},$$

where U is orthogonal and D is diagonal with eigenvalues $\lambda_1, \lambda_2, \ldots, \lambda_d$ > 0. Applying the transformation $x = m + Uy$ and noting that $|\det U| =$

1, we have

$$\int \cdots \int \exp\left\{-\frac{1}{2}{}^t(x-m)V^{-1}(x-m)\right\} dx_1 \cdots dx_d$$

$$= \int \cdots \int \exp\left\{-\frac{1}{2}{}^ty^tUV^{-1}Uy\right\} dy_1 \cdots dy_d$$

$$= \int \cdots \int \exp\left\{-\frac{1}{2}{}^tyU^{-1}V^{-1}Uy\right\} dy_1 \cdots dy_d$$

$$(\text{by } {}^tU = U^{-1})$$

$$= \int \cdots \int \exp\left\{-\frac{1}{2}{}^ty(U^{-1}VU)^{-1}y\right\} dy_1 \cdots dy_d$$

$$= \int \cdots \int \exp\left\{-\frac{1}{2}{}^tyD^{-1}y\right\} dy_1 \cdots dy_d$$

$$= \int \cdots \int \exp\left\{-\sum_i \frac{y_i^2}{2\lambda_i}\right\} dy_1 \cdots dy_d$$

$$= \prod_i \int_{\mathbb{R}^1} \exp\left\{-\frac{y_i^2}{2\lambda_i}\right\} dy_i = \prod_i \sqrt{2\pi\lambda_i} = (2\pi)^{d/2}\sqrt{\prod_i \lambda_i}$$

$$= (2\pi)^{d/2}\sqrt{\det D} = (2\pi)^{d/2}\sqrt{\det V}.$$

This proves that the integral of f over $\mathbb{R}^1$ equals 1.

Let us mention the *multinomial distribution*, which is given by

$$P_{p_1, p_2, \ldots, p_d, k} = \left(\begin{array}{c} (k_1, k_2, \ldots, k_d) \\ \dfrac{k!}{k_1!k_2!\ldots k_d!} p_1^{k_1}p_2^{k_2}\ldots p_2^{k_d} \end{array}\right)_{k_i=1,2,\ldots,\Sigma_{i=1}^d k_i=k}$$

where $\{p_1, p_2, \ldots, p_d\}$ is a set of positive numbers with sum 1 and k is a natural number. The probability assigned to a point $(k_1, k_2, \ldots, k_d)$ is the coefficient of $t_1^{k_1}t_2^{k_2}\ldots t_d^{k_d}$ in the power series expansion of $(p_1t_1 + p_2t_2 + \cdots + p_dt_d)^k$. The term "multinomial distribution" comes from this fact. Setting $t_1 = t_2 = \cdots = t_d = 1$, we can check that the total probability equals 1.

Let $\mathscr{P}^d$ denote the set of all d-dimensional distributions. The distribution function of $\mu \in \mathscr{P}^d$ is defined by

$$F(x_1, x_2, \ldots, x_d) = \mu((-\infty, x_1] \times (-\infty, x_2] \times \cdots \times (-\infty, x_d]).$$

We define the weak topology, convergence, and convolutions in $\mathscr{P}^d$ and develop the theory on these concepts in the same way as in the one-dimensional case. Also the characteristic function $\varphi(z)$ of $\mu \in \mathscr{P}^d$ is defined by

$$\varphi(z) = \int_{\mathbb{R}^d} e^{i(x,z)} \mu(dx) \qquad (z \in \mathbb{R}^d; (x,z): \text{inner product})$$

and all the results in one dimension can be extended to the d-dimensional case without any essential change.

Exercise 2.8.

(i) The characteristic function of the Gauss distribution $N_{m,v}$ is given by

$$\varphi(z) = \exp\{i(m,z) - \tfrac{1}{2}{}^t z V z\}.$$

[*Hint*: Apply the transformation $x = m + Uy$, where U is an orthogonal matrix that diagonalizes V.]

(ii) The $\varphi(z)$ in (i) is the characteristic function of a d-dimensional distribution, even if V is just positive-definite (not necessarily strictly positive-definite). This distribution is also called *Gauss distribution $N_{m,v}$ (in the extended sense)*. [*Hint*: Since $V_n = V + n^{-1}I$ (where I is the unit matrix) is strictly positive-definite, $\mu_n = N_{m,V_n}$ is meaningful. Let $\varphi_n = \mathscr{F}\mu_n$. Then

$$\varphi_n(z) = \exp\{i(m,z) - \tfrac{1}{2}{}^t z V_n z\}$$

$$\underset{(n \to \infty)}{\to} \varphi(z) = \exp\{i(m,z) - \tfrac{1}{2}{}^t z V z\}.]$$

Also this convergence is uniform near $z = 0$. Now use Lévy's convergence theorem (*d-dimensional version*).

(iii) Use characteristic functions to prove that

$$N_{m_1,v_1} * N_{m_2,v_2} = N_{m_1+m_2,v_1+v_2}.$$

(iv) Prove that the characteristic function of the multinomial distribution $P_{p_1,p_2,\ldots,p_d,k}$ is $(p_1 e^{iz_1} + p_2 e^{iz_2} + \cdots + p_d e^{iz_d})^k$.

(v) Use (iv) to prove

$$P_{p_1,p_2,\ldots,p_d,k} * P_{p_1,p_2,\ldots,p_d,r} = P_{p_1,p_2,\ldots,p_d,k+r}.$$

2.9 Infinite-dimensional distributions

As we mentioned in Section 2.5, the infinite Cartesian product

$$\mathbb{R}^\infty = \mathbb{R}^1 \times \mathbb{R}^1 \times \cdots$$

is called the sequence space and the product topology on $\mathbb{R}^\infty$ is given by the metric

$$\rho((x_n),(y_n)) = \sum_{n=1}^{\infty} 2^{-n}[|x_n - y_n| \wedge 1],$$

with respect to which $\mathbb{R}^\infty$ is a complete separable metric space. Hence every regular probability measure on $\mathbb{R}^\infty$ is standard. We call a regular probability measure on $\mathbb{R}^\infty$ simply a distribution on $\mathbb{R}^\infty$ or an *infinite-dimensional distribution*. The set of all distributions on $\mathbb{R}^\infty$ is denoted by $\mathscr{P}^\infty$ similarly to the finite-dimensional case. Also the Borel system $\mathscr{B}(\mathbb{R}^\infty)$ coincides with the product σ-algebra $\mathscr{B}^1 \times \mathscr{B}^1 \times \cdots$ as in the finite-dimensional case.

We define a map p_n from $\mathbb{R}^\infty$ to $\mathbb{R}^n$ and a map $p_{n,m}$ $(n < m)$ from $\mathbb{R}^m$ to $\mathbb{R}^n$ by

$$p_n : (x_1, x_2, \ldots) \mapsto (x_1, x_2, \ldots, x_n)$$

and

$$p_{n,m} : (x_1, x_2, \ldots, x_m) \mapsto (x_1, x_2, \ldots, x_n).$$

Both p_n and $p_{n,m}$ are clearly continuous and so Borel-measurable. Also the following relations are obvious:

$$p_{n,m} \circ p_{m,l} = p_{n,l},$$
$$p_{n,m} \circ p_m = p_n.$$

Let μ be a distribution on $\mathbb{R}^\infty$. Since both $\mathbb{R}^\infty$ and $\mathbb{R}^n$ are complete separable metric spaces, the image measure $\mu_n = \mu p_n^{-1}$ is a regular probability measure, that is, a distribution on $\mathbb{R}^n$ by Lemma 2.4.3. Since

$$p_n^{-1}(E) = (p_{n,m} \circ p_m)^{-1}(E) = p_m^{-1}(p_{n,m}^{-1}(E)), \qquad n < m,$$

the family $\{\mu_n\}$ satisfies the so-called *Kolmogorov consistency condition*:

(K) $\qquad \mu_n = \mu_m p_{n,m}^{-1}, \qquad n < m.$

The following important theorem is the converse of this fact.

Theorem 2.9.1 (Kolmogorov's extension theorem). If $\mu_n \in \mathscr{P}^n$, $n = 1, 2, \ldots,$ satisfy the Kolmogorov consistency condition, then there exists one and only one distribution $\mu \in \mathscr{P}^\infty$ such that

$$\mu_n = \mu p_n^{-1}, \qquad n = 1, 2, \ldots.$$

Proof: We consider the following classes of subsets of $\mathbb{R}^\infty$

$$\tilde{\mathscr{B}}_n = \{\, p_n^{-1}(E_n) \mid E_n \in \mathscr{B}^n \,\}, \qquad n = 1, 2, \ldots,$$

$$\mathscr{A} = \bigcup_n \tilde{\mathscr{B}}_n \qquad (\bigcup : \text{set-theoretical union}).$$

Note that $\mathscr{A}$ is the class of all subsets of $\mathbb{R}^\infty$ that belong to at least one of $\tilde{\mathscr{B}}_n$, $n = 1, 2, \ldots$. It is obvious that $\tilde{\mathscr{B}}_n$ is a σ-algebra on $\mathbb{R}^\infty$ for every n, so $\mathscr{B}$ is an algebra (but not a σ-algebra) on $\mathbb{R}^\infty$. Since $p_n(p_n^{-1}(E_n)) = E_n$ by the surjectivity of p_n, we can represent a set $B_n \in \tilde{\mathscr{B}}_n$ as $B = p_n^{-1}(E_n)$ by a unique set $E_n \in \mathscr{B}^n$. If $m > n$, every set $B \in \tilde{\mathscr{B}}_n$ belongs to $\tilde{\mathscr{B}}_m$ because

$$B = (p_{n,m} \circ p_m)^{-1}(E) = p_m^{-1}(p_{n,m}^{-1}(E)).$$

Hence we have

$$\tilde{\mathscr{B}}_1 \subset \tilde{\mathscr{B}}_2 \subset \cdots \subset \mathscr{A} \subset \mathscr{B}^\infty.$$

Since $\mathscr{B}^\infty = \mathscr{B}^1 \times \mathscr{B}^1 \times \cdots$, we have

$$\mathscr{B}^\infty = \sigma[\mathscr{A}].$$

Since the representation of $B \in \tilde{B}_n$ in the form $B = p_n^{-1}(E_n)(E_n \in \mathscr{B}^n)$ is unique,

$$\tilde{\mu}_n(B) = \mu_n(E_n)$$

defines a probability measure on $\mathbb{R}^\infty$ with domain $\tilde{\mathscr{B}}_n$. Since every element $B = p_n^{-1}(E_n)$ $(E_n \in \mathscr{B}^n)$ is also represented as

$$B = p_m^{-1}(p_{n,m}^{-1}(E_n)), \qquad m > n,$$

we obtain

$$\tilde{\mu}_m(B) = \mu_m(p_{n,m}^{-1}(E_n)).$$

$$= \mu_n(E_n) \qquad \text{(consistency condition)}$$

$$= \tilde{\mu}_n(B).$$

This implies that $\tilde{\mu}_m$ is an extension of $\tilde{\mu}_n$ for every $m > n$. Hence we can define $\tilde{\mu}(B)$ for $B \in \mathscr{A} = \bigcup_n \tilde{\mathscr{B}}_n$ by

$$\tilde{\mu}(B) = \tilde{\mu}_n(B), \qquad B \in \tilde{\mathscr{B}}_n.$$

It is obvious that the set function $\tilde{\mu}$ is an elementary probability measure on $\mathscr{A}$. If μ and μ' are two probability measures satisfying the condition mentioned in our theorem, then the argument above shows that $\mu = \mu'$ on $\mathscr{A}$. Hence the coincidence theorem (Theorem 2.2.2) ensures that $\mu = \mu'$ on $\sigma[\mathscr{A}] = \mathscr{B}^\infty$. Since both μ and μ' are regular, μ and μ' coincide with each other completely.

Now we will prove the existence of μ. Since μ must be the extension of the elementary probability measure $\tilde{\mu}$ constructed above, it suffices to check that $\tilde{\mu}$ satisfies one of the conditions mentioned in the extension theorem of probability measures (Theorem 2.2.4):

$$A_n \in \mathscr{A}\,(n = 1, 2, \ldots),\, A_1 \supset A_2 \supset \cdots,\, a = \inf \tilde{\mu}(A_n) > 0 \Rightarrow \bigcap_n A_n \neq \varnothing.$$

Though every A_n belongs to some $\tilde{\mathscr{B}}_{k(n)}$ where $k(1) < k(2) < \cdots$, we can assume that

$$A_n \in \tilde{\mathscr{B}}_n, \qquad n = 1, 2, \ldots$$

by replacing $\{A_n\}$ by the sequence $\{A'_n\}$ defined by $A'_n = A_n\,(k(n-1) < m \leq k(n))$ if necessary. Thus A_n and $\mu(A_n)$ are represented as follows:

$$A_n = p_n^{-1}(E_n), \qquad E_n \in \mathscr{B}^n,\, \tilde{\mu}(A_n) = \mu_n(E_n);\, n = 1, 2, \ldots$$

Since μ_n is K-regular, we can find a compact set $K_n \subset E_n$ such that

$$\mu_n(E_n - K_n) < 2^{-n-1}a, \qquad n = 1, 2, \ldots.$$

Setting $C_n = p_n^{-1}(K_n)$, we have

$$\tilde{\mu}(C_n) = \mu_n(K_n), \qquad \tilde{\mu}(A_n - C_n) = \mu_n(E_n - K_n) < 2^{-n-1}a.$$

Setting $D_n = \bigcap_{i=1}^n C_i$ and noting the $\{A_n\}$ is decreasing, we obtain

$$\tilde{\mu}(A_n - D_n) = \tilde{\mu}\left(\bigcap_{i=1}^n A_i - \bigcap_{i=1}^n C_i \right) \leq \sum_{i=1}^n \tilde{\mu}(A_i - C_i)$$

$$< \sum_{i=1}^n 2^{-i-1}a < a/2,$$

which implies that

$$\tilde{\mu}(D_n) = \tilde{\mu}(A_n) - \tilde{\mu}(A_n - D_n) \geq a - a/2 > 0,$$

so

$$D_n \neq \varnothing, \qquad n = 1, 2, \ldots.$$

Take a point $\xi_n = (x_n^1, x_n^2, \ldots)$ from each D_n. Then

$$\xi_n \in D_n \subset D_1 \subset C_1,$$

so $x_n^1 = p_1(\xi_n) \in K_1,\, n = 1, 2, \ldots$. Since K_1 is compact, $\{x_n^1\}$ has a convergent subsequence $x_{k(1,1)}^1, x_{k(1,2)}^1, x_{k(1,3)}^1, \ldots$. Since $\xi_{k(1,n)} \in D_{k(1,n)} \subset D_2 \subset C_2,\, n = 2, 3, 4, \ldots$, we have

$$\left(x_{k(1,n)}^1, x_{k(1,n)}^2 \right) = p_2(\xi_{k(1,n)}) \in K_2, \qquad n = 2, 3, \ldots.$$

Hence this sequence of two-dimensional points has a convergent subsequence

$$\left(x^1_{k(2,n)}, x^2_{k(2,n)} \right), \qquad n = 1, 2, \ldots .$$

Repeating this, we obtain a convergent sequence of m-dimensional points:

$$\left(x^1_{k(m,n)}, x^2_{k(m,n)}, \ldots, x^m_{k(m,n)} \right), \qquad n = 1, 2, \ldots$$

for every m. If $m' > m$, then $\{k(m', n), n = 1, 2, \ldots\}$ is a subsequence of $\{k(m, n), n = 1, 2, \ldots\}$. Set

$$\xi_{k(n,n)} = \left(\tilde{x}^1_n, \tilde{x}^2_n, \ldots \right).$$

For each m,

$$\left(\tilde{x}^1_n, \tilde{x}^2_n, \ldots, \tilde{x}^m_n \right), \qquad n = m+1, m+2, \ldots,$$

is convergent, being a subsequence of the convergent sequence of n-dimensional points obtained above. Set

$$x^m = \lim_{n \to \infty} \tilde{x}^m_n, \qquad m = 1, 2, \ldots .$$

Then

$$\left(\tilde{x}^1_n, \tilde{x}^2_n, \ldots, \tilde{x}^m_n \right) \to \left(x^1, x^2, \ldots, x^m \right) \qquad (n \to \infty), m = 1, 2, \ldots .$$

If $k(n, n) > m$, then

$$\left(\tilde{x}^1_n, \tilde{x}^2_n, \ldots, \tilde{x}^m_n \right) = p_m\left(\xi_{k(n,n)} \right) \in p_m\left(D_{k(n,n)} \right) \subset p_m(D_m)$$

$$\subset p_m(C_m) = K_m.$$

Since K_m is compact,

$$\left(x^1, x^2, \ldots, x^m \right) \in K_m, \qquad m = 1, 2, \ldots .$$

This implies that

$$\xi = \left(x^1, x^2, \ldots \right) \in p_m^{-1}(K_m) = C_m, \qquad m = 1, 2, \ldots,$$

proving that

$$\bigcap_m A_m \supset \bigcap_m C_m \neq \varnothing. \qquad \blacksquare$$

Exercise 2.9. Use Kolmogorov's extension theorem to define the direct product measure of a sequence of one-dimensional distributions μ_n, $n = 1, 2, \ldots$. [*Hint:* Check that the sequence $\nu_n = \mu_1 \times \mu_2 \times \cdots \times \mu_n \in \mathscr{P}^n$ $(n = 1, 2, \ldots)$ satisfies the Kolmogorov consistency condition.]

3

Fundamental concepts in probability theory

The aim of this chapter is to rigorously define fundamental concepts in probability theory for general trials in terms of measure theory. The base probability space is always denoted by (Ω, P) and a general element of Ω by ω. Intuitively, Ω is the sample space for the trial in consideration, ω denotes a generic sample point, and $P(A)$ represents the probability that the sample point observed in the trial is in A. In this book we assume (Ω, P) to be a *separable perfect probability measure space* to establish a more natural theory; separability and perfectness of a measure space will be defined in Section 3.1.

The idea of establishing probability theory in terms of measure theory can be found in the proof of the strong law of large numbers by E. Borel (1909) and in the study of Brownian motion by N. Wiener (1920–24). However, it was A. Kolmogorov (1933) that showed that all fields of probability theory can be discussed by this idea. At the beginning, Kolmogorov assumed only completeness of the probability measure P, but he later proposed to assume its perfectness in connection with the definition of the probability distribution of a random variable. In this book we propose an additional assumption: "separability of P." Similar proposals have been made by G. Mackey, D. Blackwell, L. Schwartz, P. Cartier, and others.

3.1 Separable perfect probability measures

Let μ be a complete probability measure on a set S. μ is called *perfect* if for every μ-measurable map $f: S \to \mathbb{R}^1$ the image measure μf^{-1} is regular. In this case the probability space (S, μ) is also called perfect. For example, every regular probability measure on a complete separable metric space is perfect by virtue of Lemma 2.4.3.

Let $\mathscr{A}$ be a class of subsets of S. $\mathscr{A}$ is called a *separating class* on S (or $\mathscr{A}$ is said to separate any two points in S), if for every $s_1, s_2 \in S$ ($s_1 \neq s_2$)

110

$\mathscr{A}$ contains a set A such that

$$1_A(s_1) \neq 1_A(s_2).$$

It is obvious that if $\mathscr{A}$ is a separating class on S, so is $\sigma[\mathscr{A}]$. The converse is also true. Suppose that $\mathscr{A}$ is not a separating class. Then we can find $s_1, s_2 \in S$ ($s_1 \neq s_2$) such that

$$1_A(s_1) = 1_A(s_2)$$

for every $A \in \mathscr{A}$. Let $\mathscr{B}$ denote the class of all sets $A \subset S$ with this property. Then $\mathscr{B}$ includes $\mathscr{A}$ obviously and is a σ-algebra on S by the following equalities:

$$1_S = 1, \qquad 1_{A^c} = 1 - 1_A, \qquad 1_{\cap A_n} = \prod 1_{A_n}.$$

Hence $\mathscr{B}$ includes $\sigma[\mathscr{A}]$. This implies that $\sigma[\mathscr{A}]$ is not separating. Thus, if $\sigma[\mathscr{A}]$ is separating, so is $\mathscr{A}$.

Let μ be a complete probability measure on a set S. μ is called *separable* if its domain $\mathscr{D}(\mu)$ includes a separating countable class. In this case, the probability space (S, μ) is also called separable. For example, every regular probability measure on $\mathbb{R}^n$ is separable for every $n = 1, 2, \ldots, \infty$.

Lemma 3.1. Let T be a subset of $\mathbb{R}^1$ and let i denote the *inclusion map* from T into $\mathbb{R}^1$, that is, the map $i: T \to \mathbb{R}^1$ such that $i(x) = x$ for every $x \in T$. Let μ be a complete probability measure on T and let ν denote the image measure μi^{-1} on $\mathbb{R}^1$. Then ν is regular if and only if μ is K-regular.

Proof: Since ν is a complete probability measure on $\mathbb{R}^1$, ν is regular if and only if ν is K-regular. Hence our theorem claims that the K-regularity of μ is equivalent to that of ν. But we can easily check this fact by noting that

$$T \in \mathscr{D}(\nu), \qquad \nu(T) = 1, \qquad \mu = \nu|_T. \qquad \blacksquare$$

Theorem 3.1.1. A complete measure is perfect and separable if and only if it is strictly isomorphic to a K-regular probability measure on a subset of $\mathbb{R}^1$.

Proof: Let μ be a complete probability measure on a set S. Suppose that μ is perfect and separable. Then $\mathscr{D}(\mu)$ includes a separating countable class $\mathscr{A} = \{A_1, A_2, \dots\}$. Define a map $f: S \to \mathbb{R}^1$ by

$$f(s) = \sum_{n=1}^{\infty} 3^{-n} e_n(s), \qquad e_n = 1_{A_n}, \quad n = 1, 2, \dots.$$

As $e_n: S \to \mathbb{R}^1$ is μ-measurable, so is $f: S \to \mathbb{R}^1$. Since μ is perfect, then the image measure $\nu = \mu f^{-1}$ is a regular probability measure on $\mathbb{R}^1$. Since $\mathscr{A}$ is separating, f is injective. If we restrict the range of f to the image $T = f(S)(\subset \mathbb{R}^1)$ we obtain a bijective map $g: S \to T$. Then the image measure $\theta = \mu g^{-1}$ is strictly isomorphic to μ. Denoting the inclusion map from T into $\mathbb{R}^1$ by $i: T \to \mathbb{R}^1$, we obtain

$$f = i \circ g,$$
$$\nu = \mu f^{-1} = (\mu g^{-1}) i^{-1} = \theta i^{-1}.$$

Since ν is a regular probability measure on $\mathbb{R}^1$, θ is K-regular by the lemma proved above. This proves the "only if" part.

To prove the "if" part it suffices to show that every K-regular probability measure μ on a set $S \subset \mathbb{R}^1$ is separable and perfect. Since $\mathscr{D}(\mu)$ includes a separating class $\mathscr{A} = \{(-\infty, r) \cap S | r \in Q\}$, μ is separable. Let $f: S \to \mathbb{R}^1$ be μ-measurable and let i denote the inclusion map from S into $\mathbb{R}^1$. Then the image measure $\mu_1 = \mu i^{-1}$ is a regular probability measure on $\mathbb{R}^1$ by the lemma proved above. Define $f_1: \mathbb{R}^1 \to \mathbb{R}^1$ by

$$f_1(x) = f(x) \text{ on } S \qquad \text{and} \qquad f_1(x) = 0 \text{ on } \mathbb{R}^1 - S.$$

Since

$$\mu_1(\mathbb{R}^1 - S) = \mu(i^{-1}(\mathbb{R}^1 - S)) = \mu(\varnothing) = 0,$$

$f_1: \mathbb{R}^1 \to \mathbb{R}^1$ is μ-measurable and

$$\mu_1 f_1^{-1} = \mu f^{-1},$$

Since μ_1 is a regular probability measure on $\mathbb{R}^1$, $\mu_1 f_1^{-1}$ (i.e., μf^{-1}) is also regular, by Lemma 2.4.3. This proves that μ is perfect. ∎

Theorem 3.1.2. Let μ be a separable perfect probability measure on a set S. If $\mathscr{A} = \{A_1, A_2, \dots\} \subset \mathscr{D}(\mu)$ is a separating class on S, then μ is the Lebesgue extension of its restriction to $\sigma[A]$.

Proof: Let f be the injective map from S into $\mathbb{R}^1$ that was introduced in the beginning of the proof of the last theorem. Since f is μ-measurable and μ is perfect, the image measure $\nu = \mu f^{-1}$ on $\mathbb{R}^1$ is regular. Hence ν is the Lebesgue extension of its restriction to $\mathscr{B}^1$. Since $f^{-1}(f(A)) = A$ for every A by the injectivity of f, we obtain

$$A \in D(\mu) \Rightarrow f(A) \in D(\nu) \qquad \text{and} \qquad \mu(A) = \nu(f(A))$$

$$\Rightarrow B_1 \subset f(A) \subset B_2, \qquad \nu(B_2 - B_1) = 0 \quad \text{for some } B_1, B_2 \in \mathscr{B}^1$$

$$\Rightarrow B_1' \subset A \subset B_2', \qquad \mu(B_2' - B_1') = 0 \quad \text{for some } B_1', B_2' \in f^{-1}(\mathscr{B}^1).$$

To complete the proof it suffices to check that

$$\sigma[\mathscr{A}] \supset f^{-1}(\mathscr{B}^1).$$

As 1_{A_n} is measurable $\sigma[\mathscr{A}]/\mathscr{B}^1$ by $A_n \in \mathscr{A} \subset \sigma[A]$, so is $f \equiv \sum_n 3^{-n} 1_{A_n}$. This implies the inclusion relation above. ∎

Theorem 3.1.3. Let S and T be any sets and let f be a map from S into T. If μ is a perfect probability measure on S, then the image measure $\nu = \mu f^{-1}$ is also a perfect probability measure on T.

Proof: Let $g \colon T \to \mathbb{R}^1$ be ν-measurable. Then $h = g \circ f \colon S \to \mathbb{R}^1$ is μ-measurable, because

$$\Gamma \in \mathscr{B}^1 \Rightarrow g^{-1}(\Gamma) \in \mathscr{D}(\nu) \Rightarrow h^{-1}(\Gamma) = f^{-1}(g^{-1}(\Gamma)) \in \mathscr{D}(\mu).$$

Hence μh^{-1} is regular by the perfectness of μ. But

$$\mu h^{-1} = (\mu f^{-1}) g^{-1} = \nu g^{-1}$$

Therefore νg^{-1} is regular. This implies that ν is perfect. ∎

Theorem 3.1.4. Let S be a topological space such that $\mathscr{B}(S)$ is σ-generated, that is, generated by a countable class. (For example, every topological space with a countable open base has this property.) If μ is a perfect probability measure on S and if $\mathscr{D}(\mu) \supset \mathscr{B}(T)$, then μ is separable and regular.

Proof: Suppose that $\mathscr{A} = \{A_n\}$ generates $\mathscr{B}(S)$. Every singleton is closed and so belongs to $\mathscr{B}(S)$. Hence $\mathscr{B}(S)$ is a separating class on S and so is $\mathscr{A}$. Since $\mathscr{A} \subset \mathscr{B}(S) \subset \mathscr{D}(\mu)$, μ is separable. Also, μ is perfect by the assumption. Therefore, μ is the Lebesgue extension of its restriction to $\sigma[\mathscr{A}]$ by Theorem 3.1.2. Since $\sigma[A] = \mathscr{B}(S)$, μ is regular. ∎

Exercise 3.1.

    (i) Prove that every regular probability measure on a complete separable metric space is separable and perfect. [*Hint:* Use

Lemma 2.4.3.]

(ii) Suppose that μ is a complete probability measure on a complete separable metric space such that $\mathscr{D}(\mu) \supset \mathscr{B}(T)$. Prove that μ is perfect if and only if μ is regular. [*Hint*: Regularity implies perfectness by (i), and the converse holds by Theorem 3.1.4.]

(iii) Let μ be a perfect probability measure and T a topological space such that $\mathscr{B}(T)$ is σ-generated. Prove that if $f: S \to T$ is μ-measurable, that is, measurable $\mathscr{D}(\mu)/\mathscr{B}(T)$, then the image measure $\nu = \mu f^{-1}$ is separable, perfect and regular. [*Hint*: $\mathscr{D}(\nu) \supset \mathscr{B}(T)$ by ν-measurability of f. Now use Theorems 3.1.3 and 3.1.4.]

(iv) Prove that every standard probability measure is perfect. [*Hint*: Use Theorem 3.1.1 and Lemma 3.1.]

(v) Let μ be a complete probability measure on a set S. Prove that μ is separable and perfect if and only if μ is standard and S has power $\leqq \mathbf{c}$.

3.2 Events and random variables

Let (Ω, P) be a separable perfect probability space that represents the pair of the sample space and the probability measure for the trial in consideration. Let $\alpha = \alpha(\omega)$ be a condition concerning a generic element ω in Ω. The set of all values of ω for which $\alpha(\omega)$ holds is called the *extension* of α, written $\{\omega | \alpha(\omega)\}$, $\{\alpha(\omega)\}$, or $\{\alpha\}$. In case $\{\alpha\}$ is P-measurable, α is called an *event* and $P\{\alpha\}$ is called the probability of occurrence of α. If $P\{\alpha\} = 1$, then α is said to occur almost surely, written $\alpha(\omega)$ a.s.

The operations $\alpha \vee \beta$, $\alpha \wedge \beta$, and $\alpha^{\neg}$ are defined in the same way as in the finite-trial case discussed in Chapter 1. Note that if $\{\alpha\}$ and $\{\beta\}$ are P-measurable, so are

$$\{\alpha \vee \beta\} = \{\alpha\} \cup \{\beta\}, \qquad \{\alpha \wedge \beta\} = \{\alpha\} \cap \{\beta\}, \qquad \{\alpha^{\neg}\} = \{\alpha\}^{c}.$$

Similarly, we can define countable operations

$$\alpha_1 \vee \alpha_2 \vee \cdots \qquad \left(\text{also written } \bigvee_n \alpha_n \text{ or } (\exists n)\alpha_n\right)$$

and

$$\alpha_1 \wedge \alpha_2 \wedge \cdots \qquad \left(\text{also written } \bigwedge_n \alpha_n \text{ or } (\forall n)\alpha_n\right).$$

If $\{\alpha_n\} \in \mathscr{D}(P)$ $(n = 1, 2, \ldots)$, then

$$\left\{\bigvee_n \alpha_n\right\} = \bigcup_n \{\alpha_n\} \in \mathscr{D}(P), \qquad \left\{\bigwedge_n \alpha_n\right\} = \bigcap_n \{\alpha_n\} \in \mathscr{D}(P).$$

For a given sequence of events $\alpha_1(\omega), \alpha_2(\omega), \ldots$, the event that an infinite number of events among these events occur and the event that all but a finite number of events among these events occur are denoted by

$$\alpha_n \text{ i.o.} \qquad (\text{i.o.} = \text{infinitely often})$$

and

$$\alpha_n \text{ f.e.} \qquad (\text{f.e.} = \text{with a finite number of exceptions}),$$

respectively. Note that if $\{\alpha_n\} \in \mathscr{D}(P)$ $(n = 1, 2, \ldots)$, then

$$\{\alpha_n \text{ i.o.}\} = \left\{\bigwedge_n \bigvee_{k > n} \alpha_k\right\} = \bigcap_n \bigcup_{k > n} \{\alpha_k\} \in \mathscr{D}(P)$$

and

$$\{\alpha_n \text{ f.e.}\} = \left\{\bigvee_n \bigwedge_{k > n} \alpha_k\right\} = \bigcup_n \bigcap_{k > n} \{\alpha_k\} \in \mathscr{D}(P).$$

Hence $\{\alpha_n \text{ i.o.}\}$ and $\{\alpha_n \text{ f.e.}\}$ are equal to the upper and lower limits of the sequence $\{\alpha_n\}$, $n = 1, 2, \ldots$, respectively.

All the properties of probability are reduced to those of probability measure and hence need not be listed here. But the following fact, which is not emphasized in textbooks on measure theory, is very important in probability theory.

Theorem 3.2.1 (The Borel–Cantelli lemma).

$$\sum_{n=1}^{\infty} P\{\alpha_n\} < \infty \Rightarrow P(\alpha_n \text{ i.o.}) = 0, \qquad P(\alpha_n^- \text{ f.e.}) = 1.$$

Proof: Let A_n denote $\{\alpha_n\}$. Then the theorem is rephrased as follows:

$$\sum_{n=1}^{\infty} P(A_n) < \infty \Rightarrow P\left(\limsup_{n \to \infty} A_n\right) = 0, \qquad P\left(\liminf_{n \to \infty} A_n^c\right) = 1.$$

Denoting the upper and lower limits above by S and L, respectively, we obtain $L = S^c$ and

$$P(S) \leq P\left(\bigcup_{k > n} A_k\right) \leq \sum_{k > n} P(A_k)$$

for every n. Since the right-hand side tends to 0 as $n \to \infty$ by the assumption of our theorem, we have $P(S) = 0$, and so $P(L) = P(S^c) = 1$.

∎

Now we will define real random variables. Let $X(\omega)$ be a real-valued function on (Ω, P). Then the condition $X(\omega) \leq a$ is regarded as a condition concerning ω. It can be called an event if and only if its extension

$$\{ X(\omega) \leq a \} = X^{-1}(-\infty, a]$$

is P-measurable, namely, if $X(\omega)$ is P-measurable. Keeping this observation in mind, we call a P-measurable real-valued function on (Ω, P) a *real random variable*.

Let $X(\omega)$ be a real random variable. Every condition on $X(\omega)$ is equivalent to a condition of the form $X(\omega) \in E$ ($E \subset \mathbb{R}^1$), and this condition is also regarded as a condition on ω whose extension is $X^{-1}(E)$. Hence $X(\omega) \in E$ is an event with probability $P(X^{-1}(E))$ if and only if $X^{-1}(E) \in \mathscr{D}(P)$. $P(X^{-1}(E))$ is a probability measure on $\mathbb{R}^1$ as a function of E. Keeping this observation in mind the *probability distribution* of $X(\omega)$, written P^X, is defined by

$$\mathscr{D}(P^X) = \{ E | X^{-1}(E) \in \mathscr{D}(P) \},$$

$$P^X(E) = P(X^{-1}(E)).$$

In other words, P^X is defined to be the image measure PX^{-1}. By virtue of perfectness of P, P^X turns out to be a regular probability measure.

Suppose that we assumed only completeness but not perfectness of P. Then P^X would be a complete extension of its restriction to $\mathscr{B}^1$. Hence P^X would be an extension of a regular probability measure on $\mathbb{R}^1$. Then P^X would not be determined by giving its distribution function $F^X(x) = P\{ X(\omega) \leq x \}$. In other words, there would be an event on (Ω, P) whose occurrence or nonoccurrence depended only on the value of $X(\omega)$ and whose probability could not be determined by giving $P\{ X(\omega) \leq x \}$ ($x \in \mathbb{R}^1$). To avoid this annoying situation, Kolmogorov proposed to assume perfectness of P.

Generalizing the notion of real random variables we define an *n-dimensional random variable* to be an $\mathbb{R}^n$-valued function on Ω that is P-measurable, that is, measurable $\mathscr{D}(P)/\mathscr{B}^n$, where $n = 1, 2, \ldots, \infty$. The *probability distribution* of an n-dimensional random variable $X(\omega)$, written P^X, is defined to be the image measure PX^{-1}, as in the one-dimensional case. Since $\mathbb{R}^n$ has a countable open base, we can apply Exercise 3.1(iii) to check that P^X is a regular probability measure on $\mathbb{R}^n$.

More generally, in case T is a topological space such that $\mathscr{B}(T)$ is generated by a countable class, we can also define a T-valued random variable to be a T-valued function on Ω that is measurable $\mathscr{D}(P)/\mathscr{B}(T)$. The *probability distribution* of a T-valued random variable $X(\omega)$, written P^X, is defined to be the image measure PX^{-1}. P^X is a regular probability measure on T by Exercise 3.1(iii).

Similarly, we can define a random variable with values in a general topological space, but its probability law is not always regular. Since such a random variable never appears in practical applications, we will not consider it in this book.

We can also consider a random variable with values in an abstract set S. Let $X(\omega)$ be an S-valued function defined on Ω. Since there is no Borel system on S, P-measurability of $X(\omega)$ is meaningless. We will introduce P-separability instead of P-measurability. An S-valued function $X(\omega)$ is called *P-separable* if there exists a countable separating class $\mathscr{A}$ such that

$$X^{-1}(A) \in \mathscr{D}(P) \qquad \text{for every } A \in \mathscr{A}.$$

A P-separable S-valued function $X(\omega)$ on Ω is called an S-valued random variable, and the *probability distribution* of $X(\omega)$, P^X, is defined to be the image measure PX^{-1}. The P-separability of $X(\omega)$ implies that P^X is separable. Since P is complete, P^X is also complete. Since P is perfect, P^X is also perfect, by Theorem 3.1.3. Thus P^X is a separable perfect complete probability measure on S.

Let S be a topological space such that $\mathscr{B}(S)$ is generated by a countable class. An S-valued function $X(\omega)$ is a random variable if $X(\omega)$ is P-measurable. Also $X(\omega)$ is regarded as a random variable in the sense described just above by ignoring the topological structure of S if $X(\omega)$ is P-separable. Hence we have two different concepts of an S-valued random variable. The former is more narrow than the latter, as we can easily check. When we consider a random variable taking values in a topological space, we understand it in the former sense, unless explicitly stated otherwise. However, if we want to emphasize it, the random variable is called a *topological random variable*. A real random variable is an $\mathbb{R}^1$-valued topological random variable and an n-dimensional vector random variable (often called an n-dimensional random vector) is an $\mathbb{R}^n$-valued topological random variable.

Let us summarize what we have observed above. If T is a topological space such that $\mathscr{B}(T)$ is σ-generated, a T-valued topological random variable $X(\omega)$ is defined to be a P-measurable T-valued function. If S is a set, then an S-valued random variable $Y(\omega)$ is defined to be a P-separable S-valued function. The probability measure $P^X = PX^{-1}$ is a complete, separable, perfect, and regular probability measure on T, and the probability measure $P^Y = PY^{-1}$ is a complete, separable, and perfect probability measure on S. A T-valued topological random variable is regarded as a T-valued random variable by ignoring the topological structure in T. Hence the following facts concerning random variables hold also for topological random variables.

Let $X(\omega)$ be an S-valued random variable. Since there exists a countable separating class by P-separability of $X(\omega)$, the power of the set S, written $\#S$, must be no more than $\mathbf{c}$, the power of continuum. In fact, if we set

$$\varphi(s) = \big(e_1(s), e_2(s), \ldots\big), \qquad v_n = 1_{A_n}, \qquad n = 1, 2, \ldots,$$

then φ is an injective map from S into $\mathbf{I} = \{0,1\}^{\mathbb{N}}$, so

$$\#S = \#\varphi(S) \le \#\mathbf{I} = \mathbf{c}.$$

Next we will mention the change of the range of a random variable. Let $X(\omega)$ be an S-valued random variable and T an arbitrary set of power $\le \mathbf{c}$, which includes U. Then $X_T(\omega) = 1_{S} \circ X(\omega)$ ($i_{U,T}$ = the imbedding map from S into T) is a T-valued function on Ω. Since $X_T(\omega) = X(\omega)$ for every $\omega \in \Omega$, $X_T(\omega)$ is essentially the same as $X(\omega)$. The difference is that $X_T(\omega)$ is T-valued, whereas $X(\omega)$ is S-valued. Such change of view occurs often in analysis, for example, regarding an integer-valued function as a real-valued function. Now we will prove $X_T(\omega)$ is a T-valued random variable. Using $X(\omega)$ is an S-valued variable, we have a countable separating class in S, $\{A_n\}$ such that

$$X^{-1}(A_n) \in \mathcal{D}(P), \qquad n = 1, 2, \ldots,$$

Setting $B_n = A_n \cap T \, (U = X(\Omega))$, we obtain

$$X_T^{-1}(B_n) = X^{-1}(B_n) = X^{-1}(A_n) \in \mathcal{D}(P), \qquad n = 1, \ldots$$

As $\{A_n\}$ is a separating class on S, so is $\{B_n\}$ on U. Since $\#T \le \mathbf{c}$, $T - U$ corresponds to a subset of $\mathbb{R}^1$ in 1 to-1, countable separating class $\{C_n\}$ on $T - U$. But

$$X_T^{-1}(C_n) = \emptyset \in \mathcal{D}(P), \qquad n = 1, 2, \ldots$$

by $X_T(\Omega) = U$. Hence $\{B_n, C_n, n = 1, 2, \ldots\}$ is a countable class on T. This implies that $X_T(\omega)$ is P-separable random variable. The probability distribution of X_T separable perfect complete probability measure on T.

More generally, in case T is a topological space such that $\mathscr{B}(T)$ is generated by a countable class, we can also define a T-valued random variable to be a T-valued function on Ω that is measurable $\mathscr{D}(P)/\mathscr{B}(T)$. The *probability distribution* of a T-valued random variable $X(\omega)$, written P^X, is defined to be the image measure PX^{-1}. P^X is a regular probability measure on T by Exercise 3.1(iii).

Similarly, we can define a random variable with values in a general topological space, but its probability law is not always regular. Since such a random variable never appears in practical applications, we will not consider it in this book.

We can also consider a random variable with values in an abstract set S. Let $X(\omega)$ be an S-valued function defined on Ω. Since there is no Borel system on S, P-measurability of $X(\omega)$ is meaningless. We will introduce P-separability instead of P-measurability. An S-valued function $X(\omega)$ is called *P-separable* if there exists a countable separating class $\mathscr{A}$ such that

$$X^{-1}(A) \in \mathscr{D}(P) \qquad \text{for every } A \in \mathscr{A}.$$

A P-separable S-valued function $X(\omega)$ on Ω is called an S-valued random variable, and the *probability distribution* of $X(\omega)$, P^X, is defined to be the image measure PX^{-1}. The P-separability of $X(\omega)$ implies that P^X is separable. Since P is complete, P^X is also complete. Since P is perfect, P^X is also perfect, by Theorem 3.1.3. Thus P^X is a separable perfect complete probability measure on S.

Let S be a topological space such that $\mathscr{B}(S)$ is generated by a countable class. An S-valued function $X(\omega)$ is a random variable if $X(\omega)$ is P-measurable. Also $X(\omega)$ is regarded as a random variable in the sense described just above by ignoring the topological structure of S if $X(\omega)$ is P-separable. Hence we have two different concepts of an S-valued random variable. The former is more narrow than the latter, as we can easily check. When we consider a random variable taking values in a topological space, we understand it in the former sense, unless explicitly stated otherwise. However, if we want to emphasize it, the random variable is called a *topological random variable*. A real random variable is an $\mathbb{R}^1$-valued topological random variable and an n-dimensional vector random variable (often called an n-dimensional random vector) is an $\mathbb{R}^n$-valued topological random variable.

Let us summarize what we have observed above. If T is a topological space such that $\mathscr{B}(T)$ is σ-generated, a T-valued topological random variable $X(\omega)$ is defined to be a P-measurable T-valued function. If S is a set, then an S-valued random variable $Y(\omega)$ is defined to be a P-separable S-valued function. The probability measure $P^X = PX^{-1}$ is a complete, separable, perfect, and regular probability measure on T, and the probability measure $P^Y = PY^{-1}$ is a complete, separable, and perfect probability measure on S. A T-valued topological random variable is regarded as a T-valued random variable by ignoring the topological structure in T. Hence the following facts concerning random variables hold also for topological random variables.

Let $X(\omega)$ be an S-valued random variable. Since there exists a countable separating class by P-separability of $X(\omega)$, the power of the set S, written $\#S$, must be no more than $\mathbf{c}$, the power of continuum. In fact, if we set

$$\varphi(s) = (e_1(s), e_2(s), \ldots), \qquad e_n = 1_{A_n}, \qquad n = 1, 2, \ldots,$$

then φ is an injective map from S into $\mathbf{I} = \{0, 1\}^\infty$, so

$$\#S = \#\varphi(S) \leq \#I = \mathbf{c}.$$

Next we will mention the change of the range of a random variable. Let $X(\omega)$ be an S-valued random variable and T an arbitrary set of power $\leq \mathbf{c}$, which includes S. Then $X_T(\omega) = i_{S,T}(X(\omega))$ ($i_{S,T}$ = the inclusion map from S into T) is a T-valued function on Ω. Since $X_T(\omega) = X(\omega)$ for every $\omega \in \Omega$, $X_T(\omega)$ is essentially the same as $X(\omega)$. The only difference is that $X_T(\omega)$ is T-valued, whereas $X(\omega)$ is S-valued. Such a change of view occurs often in analysis, for example, regarding an integer-valued function as a real-valued function. Now we will prove that $X_T(\omega)$ is a T-valued random variable. Since $X(\omega)$ is an S-valued random variable, we have a countable separating class in S, $\{A_n\}$ such that

$$X^{-1}(A_n) \in \mathcal{D}(P), \qquad n = 1, 2, \ldots.$$

Setting $B_n = A_n \cap U$ ($U = X(\Omega)$), we obtain

$$X_T^{-1}(B_n) = X^{-1}(B_n) = X^{-1}(A_n) \in \mathcal{D}(P), \qquad n = 1, 2, \ldots.$$

As $\{A_n\}$ is a separating class on S, so is $\{B_n\}$ on U. Since $\#(T - U) \leq \#T \cap \mathbf{c}$, $T - U$ corresponds to a subset of $\mathbb{R}^1$ in 1-to-1, so there is a countable separating class $\{C_n\}$ on $T - U$. But

$$X_T^{-1}(C_n) = \varnothing \in \mathcal{D}(P), \qquad n = 1, 2, \ldots,$$

by $X_T(\Omega) = U$. Hence $\{B_n, C_n, n = 1, 2, \ldots\}$ is a countable separating class on T. This implies that $X_T(\omega)$ is P-separable and so a T-valued random variable. The probability distribution of X_T, $P^{X_T} = PX_T^{-1}$ is a separable perfect complete probability measure on T.

A similar observation is possible in case S and T are topological spaces, if $\mathscr{B}(S)$ and $\mathscr{B}(T)$ are σ-generated, and if $\mathscr{B}(T) \cap U = \mathscr{B}(S) \cap U$, where $\mathscr{B} \cap U = \{B \cap U | B \in \mathscr{B}\}$.

Let $X(\omega)$ be an S-valued random variable and $Y(\omega)$ a T-valued one. If the value $Y(\omega)$ is determined by the value $X(\omega)$, namely, if

$$X(\omega_1) = X(\omega_2) \Rightarrow Y(\omega_1) = Y(\omega_2),$$

then Y is called a *daughter* (*variable*) of X, and X is called a *mother* (*variable*) of Y. In this case we find a map $f: S \to T$ such that

$$f(X(\omega)) = Y(\omega); \quad \text{i.e., } f \circ X = Y.$$

If $s \in X(\Omega)$, then $Y(X^{-1}(s))$ is a singleton, which is denoted by $\{f(s)\}$. If $s \in S - X(\Omega)$, we define $f(s)$ to be any point in T. Since $Y(\omega)$ is P-separable, we have a countable separating class on T $\mathscr{A} \subset \mathscr{D}(PY^{-1})$. Since

$$PY^{-1} = (PX^{-1})f^{-1} = P^X f^{-1},$$

we have

$$\mathscr{A} \subset \mathscr{D}(P^X f^{-1}).$$

This implies that $f(s)$ is a P^X-separable function on (S, P^X); namely, that $f(s)$ is a T-valued random variable on the probability space (S, P^X).

If $Y(\omega)$ is a T-valued topological random variable, then we have

$$\mathscr{B}(T) \subset \mathscr{D}(PY^{-1}) = \mathscr{D}(P^X f^{-1})$$

so $f(s)$ turns out to be a T-valued topological random variable on (S, P^X).

The probability space (S, P^X) is called the *probability space* of X, and $f(s)$ is called a representation of $Y(\omega)$ on (S, P^X). Such a function f is unique on $X(\Omega)$ but not elsewhere. It is unique a.s., because

$$P^X(X(\Omega)) = P(X^{-1}(X(\Omega))) = P(\Omega) = 1$$

Conversely if $f(s)$ is a T-valued random variable on (S, P^X), then $Y(\omega) := f(X(\omega))$ is a T-valued random variable on (Ω, P) and is a daughter of X. Therefore, when we consider only the daughters of $X(\omega)$, we can take (S, P^X) for the base probability space and consider $f(s)$ in place of $f(X(\omega))$. We can avoid the nonuniqueness of f by changing the range of X into $S' = X(\Omega)$ and taking $(S', P^X|_{S'})$ for the probability space of X.

Let $I: \Omega \to \Omega$ be the identity map. Then $I(\omega)$ is an Ω-valued random variable on (Ω, P). Since every random variable $X(\omega)$ on (Ω, P) is

represented as

$$X(\omega) \equiv X(I(\omega)),$$

$\omega \equiv I(\omega)$ is a common mother of all random variables.

We shall define the joint variable of a countable number of random variables. Let $X_n(\omega)$ be an S_n-valued random variables for $n = 1, 2, \ldots$. Let

$$S = \prod_n S_n, \qquad p_k : S \to S_k \text{ (projection)}, \qquad k = 1, 2, \ldots.$$

Since X_n is P-separable, we have a countable separating class $A_n = \{A_{n1}, A_{n2}, \ldots\}$ such that

$$X_n^{-1}(A_{nk}) \subset \mathcal{D}(P), \qquad k = 1, 2, \ldots.$$

The S-valued function on Ω:

$$X(\omega) = (X_1(\omega), X_2(\omega), \ldots)$$

is P-separable, because

$$\mathcal{A} = \{p_n^{-1}(A_{nk}) \mid n, k = 1, 2, \ldots\}$$

is a countable separating class and

$$X^{-1}(p_n^{-1}(A_{nk})) = (p_n \circ X)^{-1}(A_{nk})$$
$$= X_n^{-1}(A_{nk}) \in \mathcal{D}(P).$$

Therefore X is an S-valued random variable, called the *joint variable* of X_n, $n = 1, 2, \ldots$, and P^X is the Lebesgue extension of $P^X|_{\sigma[\mathcal{A}]}$ by Theorem 3.1.2.

Let $X_n(\omega)$ be a topological random variable with values in a topological space S_n having a countable open base for $n = 1, 2, \ldots$. Then the topological product $S = \prod_n S_n$ has a countable open base, and the S-valued function

$$X(\omega) = (X_1(\omega), X_2(\omega), \ldots)$$

defines an S-valued topological random variable, called the *topological joint variable*. The topological joint variable of a countable number of real random variables is a random vector.

The space $\overline{\mathbb{R}} = [-\infty, \infty]$ with the usual topology is called the extended real line. It is a topological space with a countable open base. Hence we can define an $\overline{\mathbb{R}}$-valued random variable, called an *extended real random variable*. Since $\mathbb{R}^1$ is a subset of $\overline{\mathbb{R}}$ and the topology in $\mathbb{R}^1$ coincides with the one induced from that in $\overline{\mathbb{R}}$, every real random variable is regarded as an $\overline{\mathbb{R}}$-valued random variable by changing its range.

Let $X(\omega)$ be an extended real random variable. For any P-measurable set we define

$$E(X, A) = \int_A X(\omega) P(d\omega).$$

If $X(\omega) \geq 0$ on A, this is well-defined and takes a value in $[0, \infty]$. If one of $E(X^+, A)$ and $E(X^-, A)$ $(X^{\pm}(\omega) = \max(\pm X(\omega), 0))$ is finite, then we set

$$E(X, A) = E(X^+, A) - E(X^-, A).$$

$E(X, A)$ takes a value in $[-\infty, \infty]$. If $E(|X|, A) < \infty$ then $X(\omega)$ is called *integrable* on A and $E(X, A)$ takes a real value. We denote $E(X, \Omega)$ simply by EX and call it the *expectation* (or *mean value*) of X. Observing that $E(X, A) = E(X \cdot 1_A)$, we can reduce the properties of $E(X, A)$ to those of EX.

We list some basic properties of expectation, where $(\pm\infty) \cdot 0 = 0$ for convention and $EX = EY$ means that if one of these is well-defined then so is the other and both values coincide with each other.

Theorem 3.2.2 (Properties of mean values).

   (i) $E(1_A) = P(A)$.

   (ii) $X(\omega) = Y(\omega)$ a.s. $\Rightarrow EX = EY$.

   (iii) $E(aX) = aE(X)$ (a: a finite constant).

   (iv) If $X(\omega) \geq 0$ a.s., then

$$EX = 0 \Rightarrow X(\omega) = 0 \quad \text{a.s.,}$$

$$EX < \infty \Rightarrow X(\omega) < \infty \quad \text{a.s.}$$

   (v) (*The monotone convergence theorem.*) If either (1) $\{X_n(\omega)\}$ is increasing a.s. and $EX_1 > -\infty$ or (2) $\{X_n(\omega)\}$ is decreasing a.s. and $EX_1 < \infty$, then

$$E\left(\lim_{n \to \infty} X_n\right) = \lim_{n \to \infty} EX_n.$$

   (vi) (*Fatou's lemma*).

$$E\left(\inf_n X_n\right) > -\infty \Rightarrow E\left(\liminf_{n \to \infty} X_n\right) \leq \liminf_{n \to \infty} EX_n,$$

$$E\left(\sup_n X_n\right) < \infty \Rightarrow E\left(\limsup_{n \to \infty} X_n\right) \geq \limsup_{n \to \infty} EX_n,$$

$$E\left(\sup_n |X_n|\right) < \infty \Rightarrow E\left(\liminf_{n \to \infty} X_n\right) \leq \liminf_{n \to \infty} EX_n \leq \limsup_{n \to \infty} EX_n$$

$$\leq E\left(\limsup_{n \to \infty} X_n\right).$$

(vii) (*The dominated convergence theorem.*)

$$E\left(\sup_n |X_n| \right) < \infty, \quad X(\omega) = \lim_{n \to \infty} X_n(\omega) \text{ a.s. } \Rightarrow EX = \lim_{n \to \infty} EX_n.$$

(viii) (*The exchangeability with infinite sums.*)

$$X_n(\omega) \geq 0 \quad (n = 1, 2, \dots) \Rightarrow E\left(\sum_n X_n \right) = \sum_n EX_n,$$

$$E\left(\sum_n |X_n| \right)\left(= \sum_n E|X_n| \right) < \infty \Rightarrow E\left(\sum_n X_n \right) = \sum_n EX_n.$$

(ix) (*The formula on change of variables.*) Let $X(\omega)$ be an S-valued random variable on (Ω, P) and $f(s)$ an extended real random variable on (S, P^X). Then $f(X(\omega))$ is an extended real random variable on (Ω, P) and

$$E(f(X)) = E^X(f)\left(= \int_S f(s) P^X(ds) \right).$$

Proof: All these properties except (ix) follow immediately from the corresponding ones for integrals. The property (ix) holds for $f = 1_B (B \in \mathcal{D}(P^X))$ as follows:

$$E(1_B(X)) = P(X^{-1}(B)) = P^X(B) = E^X(1_B);$$

and this is so for every linear combination of such functions with positive coefficients. Taking limits and subtracting we can verify (ix) for the general case. ∎

Exercise 3.2.

(i) Suppose that $a_n > 0$ and $\sum_n a_n < \infty$. Use the Borel–Cantelli lemma to prove that if X_n is a real random variable satisfying $P(|X_n| > a_n) < a_n$ for every $n = 1, 2, \dots$, then $\sum_n X_n(\omega)$ is absolutely convergent a.s.

(ii) Prove that every regular probability measure on $\mathbb{R}^n$ is separable and perfect for $n = 1, 2, \dots, \infty$. [*Hint*: These spaces are complete separable metric spaces.]

(iii) Prove that every P-measurable function with values in a separable metric space T is regarded as a topological random variable whose probability distribution is regular and separable. [*Hint*: Since T has a countable open base, $\mathcal{B}(T)$ is σ-generated.]

3.3   Decompositions and σ-algebras

From now on the base probability space is denoted by (Ω, P) and a generic point in Ω by ω, unless stated otherwise. A class of nonempty P-measurable subsets is called a *decomposition* of Ω if these sets are disjoint and their union is Ω. In other words, a decomposition of Ω is a disjoint covering of Ω consisting of nonempty P-measurable sets. If A is a nonempty P-measurable proper subset of Ω, then $\Delta = \{A, A^c\}$ is a decomposition of Ω. $\Delta = \{\Omega\}$ is also a decomposition of Ω. If $X(\omega)$ is a random variable, then $\Delta_X = \{X^{-1}(x) | x \in X(\Omega)\}$ is a decomposition of Ω, which is called the decomposition under X.

For a countable class $\mathscr{A} = \{A_1, A_2, \ldots\} \subset \mathscr{D}(P)$ the class of all nonempty sets ξ expressible in the form

$$\xi = A_1' \cap A_2' \cap \cdots \qquad \left(A_n' = A_n \quad \text{or} \quad A_n^c\right)$$

is a decomposition of Ω, called the decomposition generated by $\mathscr{A}$, written $\Delta_{\mathscr{A}}$. A decomposition under a countable subclass of $\mathscr{D}(P)$ is called a *separable decomposition*.

Theorem 3.3.1.

(i) Every decomposition under a random variable is separable.

(ii) Every separable decomposition is a decomposition under a real random variable.

Proof: (i) Let Δ be the decomposition under an S-valued random variable X. By the definition of a random variable we can find a countable separating class $\{B_n\} \subset 2^S$ such that

$$\mathscr{A} = \{A_n = X^{-1}(B_n)\} \subset \mathscr{D}(P).$$

Then $\Delta = \Delta_{\mathscr{A}}$ is separable and so is Δ_X. (ii) Let Δ be the decomposition generated by $\mathscr{A} = \{A_1, A_2 \ldots\} \subset \mathscr{D}(P)$. Then

$$X(\omega) = \sum_{n=1}^{\infty} 3^{-n} 1_{A_n}(\omega)$$

is a real random variable. Let ξ be any element of Δ. When ω moves in ξ, $1_{A_n}(\omega)$ remains constant for every n, and so does $X(\omega)$. This implies that ξ is included in an element of Δ_X. Conversely every element of Δ_X is included in an element of Δ, because

$$1_{A_n}(\omega) = [3^n X(\omega)] - 3[3^{n-1} X(\omega)], \qquad n = 1, 2, \ldots.$$

Since both Δ and Δ_X are decompositions, we obtain $\Delta = \Delta_X$. ∎

Let Δ and Δ' be two decompositions. If every element of Δ' is included in an element of Δ, we say that Δ' is *finer* than Δ, or Δ' is a *refinement* of Δ, written

$$\Delta' \succ \Delta \qquad (\text{or } \Delta \prec \Delta')$$

In this case every element of Δ is a disjoint union of a number of elements of Δ'.

Since it is obvious that

$$\Delta \prec \Delta', \quad \Delta' \prec \Delta'' \Rightarrow \Delta \prec \Delta'',$$
$$\Delta \prec \Delta', \quad \Delta' \prec \Delta \Rightarrow \Delta' = \Delta,$$

the relation " $\prec$ " defines a semi-order in the family of all decompositions. In accordance with this semi-order we can define the *least upper bound*, $\bigvee_\lambda \Delta_\lambda$, and the *greatest lower bound*, $\bigwedge_\lambda \Delta_\lambda$, for any given family of decompositions $\{\Delta_\lambda\}_\lambda$. But these bounds do not exist in general.

Theorem 3.3.2. Let $\{\Delta_n\}$ be a sequence of decompositions.

(i) The least upper bound $\bigvee_n \Delta_n$ exists and

$$\bigvee_n \Delta_n = \left\{ \bigcap_n A_n \neq \emptyset \,\middle|\, A_n \in \Delta_n \, (n = 1, 2, \ldots) \right\}.$$

(ii) If $\{\Delta_n\}$ is increasing, namely, if $\Delta_1 \prec \Delta_2 \prec \cdots$, then

$$\bigvee_n \Delta_n = \left\{ \bigcap_n A_n \neq \emptyset \,\middle|\, A_n \in \Delta_n \, (n = 1, 2, \ldots), A_1 \subset A_2 \subset \cdots \right\}.$$

In this case the least upper bound is called the limit of $\{\Delta_n\}$, denoted by $\lim_n \Delta_n$.

(iii) If $\{\Delta_n\}$ is decreasing, namely, if $\Delta_1 \succ \Delta_2 \succ \cdots$, then the greatest lower bound $\bigwedge_n \Delta_n$ exists;

$$\bigwedge_n \Delta_n = \left\{ \bigcup_n A_n \,\middle|\, A_n \in \Delta_n \, (n = 1, 2, \ldots), A_1 \subset A_2 \subset \cdots \right\}.$$

In this case the greatest lower bound is called the limit of $\{\Delta_n\}$, denoted by $\lim_n \Delta_n$.

Proof: (i) Let Δ denote the right-hand side of the equality. It is easy to verify that Δ is a decomposition that is finer than every Δ_k because

$$\bigcap_n A_n \subset A_k \in \Delta_k, \qquad k = 1, 2, \ldots.$$

Suppose that Δ' is a decomposition such that $\Delta' \succ \Delta_k$ for every k. For every $A' \in \Delta'$ and every k, we can find $A_k \in \Delta_k$ such that $A' \subset A_k$. Then

$$A' \subset \bigcap_k A_k \in \Delta.$$

This implies that Δ' is finer than Δ. Therefore, Δ is the least upper bound of $\{\Delta_n\}$.

(ii) Since $\{\Delta_n\}$ is increasing, every sequence $\{A_n \in \Delta_n, \ n = 1, 2, \ldots\}$ with $\bigcap_n A_n \neq \emptyset$ is decreasing. Hence (ii) is a special case of (i).

(iii) Let Δ denote the right-hand side of the equality. First we check that if $\xi \cap \eta \neq \emptyset$ ($\xi, \eta \in \Delta$), then $\xi = \eta$. Since ξ and η are expressible in the form

$$\xi = \bigcup_n A_n, \qquad \eta = \bigcup_n B_n \quad (A_n, B_n \in \Delta_n, A_n \uparrow, B_n \uparrow),$$

we obtain

$$\bigcup_{m,n} A_m \cap B_n = \xi \cap \eta \neq \emptyset.$$

Hence we can find (m, n) such that $A_m \cap B_n \neq \emptyset$. Denote $\max(m, n)$ by k_0. Then for every $k > k_0$ we have

$$A_k \cap B_k \supset A_m \cap B_n \neq \emptyset,$$

which implies $A_k = B_k$ for $k > k_0$. Thus we obtain $\xi = \eta$. Let A_1 be any element of Δ_1. Since $\{\Delta_n\}$ is decreasing, we can find $A_2, A_3, \ldots$ such that

$$A_1 \subset A_2 \subset \cdots \qquad (A_n \in \Delta_n; n = 1, 2, \ldots).$$

This implies that

$$A_1 \subset \bigcup_n A_n \in \Delta,$$

namely, that every element of Δ_1 is included in an element of Δ. As Δ_1 covers Ω, so does Δ. Thus we have proved that Δ is a decomposition of Ω. Using the same argument as in the proof of (i), we can easily verify that $\Delta = \bigwedge_n \Delta_n$. ∎

Theorem 3.3.3. The least upper bound of a countable family of separable decompositions is separable.

Proof: It suffices to observe that if Δ_n is the decomposition under a countable class $\mathscr{A}_n$ for $n = 1, 2, \ldots$, then $\bigvee_n \Delta_n$ is the decomposition generated by the union of these classes. ∎

If Δ_n is the decomposition under $X_n(\omega)$, then $\bigvee_n \Delta_n$ is the decomposition under the joint variable $X(\omega) = (X_1(\omega), X_2(\omega), \ldots)$, as is clear from the definitions. As a corollary of this fact and Theorem 3.3.1, we can also derive the theorem above.

As a special case of Theorem 3.3.3 we have that the limit of an increasing sequence of separable decompositions is separable. But the limit of a decreasing sequence of separable decompositions is not always separable, as we will see in Exercise 3.4(vi).

Theorem 3.3.4. Let $X(\omega)$ and $Y(\omega)$ be random variables. Then $Y(\omega)$ is a daughter of $X(\omega)$ if and only if $\Delta_Y \prec \Delta_X$.

Proof: Obvious by the definitions. ∎

For a decomposition Δ we define a Δ-valued function X_Δ by

$$X_\Delta(\omega) = \text{the element of } \Delta \text{ containing } \omega.$$

$X_\Delta(\omega)$ is P-separable if and only if Δ is separable. In this case $X_\Delta(\omega)$ is a Δ-valued random variable, called a random variable corresponding to Δ.

Let **B** denote the family of all σ-algebras consisting of P-measurable subsets of Ω. **B** is a semi-ordered system with respect to the inclusion relation. $\mathscr{D}(P)$ is the largest σ-algebra in **B** and $\{\varnothing, \Omega\}$ is the smallest σ-algebra in **B**. Also,

$$\mathbf{2} = \{A \in \mathscr{D}(P) | P(A) = 0 \quad \text{or} \quad 1\}$$

belongs to **B**. With respect to the semi-order in **B** we can define the least upper bound and the greatest lower bound for any given family $\{B_\lambda\} \subset \mathbf{B}$, denoted by $\bigvee_\lambda \mathscr{B}_\lambda$ (or $\sup_\lambda \mathscr{B}_\lambda$) and $\bigwedge_\lambda \mathscr{B}_\lambda$ (or $\inf_\lambda \mathscr{B}_\lambda$), respectively. It is obvious that

$$\bigvee_\lambda \mathscr{B}_\lambda = \sigma\left[\bigcup_\lambda \mathscr{B}_\lambda\right], \qquad \bigwedge_\lambda \mathscr{B}_\lambda = \bigcap_\lambda \mathscr{B}_\lambda.$$

For a given sequence $\{\mathscr{B}_n\} \subset \mathbf{B}$ we define

$$\limsup_n \mathscr{B}_n = \bigwedge_n \bigvee_{k > n} \mathscr{B}_k \qquad \text{(upper limit)}$$

and

$$\liminf_n \mathscr{B}_n = \bigvee_n \bigwedge_{k > n} \mathscr{B}_k \qquad \text{(lower limit)}.$$

If these limits are equal, we denote either one by $\lim_n \mathscr{B}_n$ (limit). This limit exists if $\{B_n\}$ is monotone; and according to whether it is increasing or decreasing, we obtain

$$\lim_n \mathscr{B}_n = \bigvee_n \mathscr{B}_n \quad \text{or} \quad \bigwedge_n \mathscr{B}_n.$$

For a decomposition Δ the class of all P-measurable sets expressible as the union of a finite or infinite number of elements of Δ is a σ-algebra, denoted by $\mathscr{B}_\Delta$.

Let $\mathscr{B}$ be a σ-algebra of P-measurable sets and Δ a decomposition. If

$$\Delta \subset \mathscr{B} \subset \mathscr{B}_\Delta,$$

$\mathscr{B}$ is said to be *subordinate* to Δ. It is obvious that B_Δ is subordinate to Δ. The family of all σ-algebras subordinate to Δ is denoted by $\mathbf{B}(\Delta)$.

For a random variable $X(\omega)$ the class

$$X^{-1}(\mathscr{D}(P^X)) = \{ X^{-1}(E) | E \in \mathscr{D}(P^X) \}$$

is a σ-algebra, denoted by $\bar{\sigma}[X]$. It is obvious that

$$\Delta = \Delta_X \Rightarrow \mathscr{B}_\Delta = \bar{\sigma}[X].$$

Hence $\bar{\sigma}[X]$ is subordinate to Δ_X.

If X is a topological random variable, we introduce another σ-algebra

$$\sigma[X] = X^{-1}(\mathscr{B}(S)) = \{ X^{-1}(E) | E \in \mathscr{B}(S) \}$$

Since it is obvious that

$$\Delta_X \subset \sigma[X] \subset \bar{\sigma}[X] = \mathscr{B}_{\Delta_X},$$

both $\sigma[X]$ and $\bar{\sigma}[X]$ are subordinate to Δ_X.

Let $\mathscr{B}_1, \mathscr{B}_2 \in \mathbf{B}$. $\mathscr{B}_1$ is said to be *equivalent* to $\mathscr{B}_2$, if

$$\mathscr{B}_1 \vee \mathbf{2} = \mathscr{B}_2 \vee \mathbf{2}.$$

This means that every element of one of $\mathscr{B}_1$ and $\mathscr{B}_2$ differs from an element of the other one by a P-null set. This relation is an equivalence relation. As a stronger equivalence relation we define $\mathscr{B}_1$ to be *strictly equivalent* to $\mathscr{B}_2$ if we can find a P-null set N such that

$$\mathscr{B}_1 \cap (\Omega - N) = \mathscr{B}_2 \cap (\Omega - N),$$

where $\mathscr{B} \cap A = \{ B \cap A | B \in \mathscr{B} \}$.

Exercise 3.3.

   (i) Let $f: S \to T$ and $\mathscr{B}$ be a σ-algebra on T. Prove that $f^{-1}(\mathscr{B})$ is a σ-algebra on S.
   (ii) Let $f: S \to T$ and $\mathscr{B}$ be a σ-algebra on S. Prove that

   $$f[\mathscr{B}] = \{ B \subset T | f^{-1}(B) \in \mathscr{B} \}$$

   is a σ-algebra on T. Give a counterexample to show that

   $$f(\mathscr{B}) = \{ f(B) | B \in \mathscr{B} \}$$

   is not always a σ-algebra on T.
   (iii) Let S_λ, $\lambda \in \Lambda$, be topological spaces, S their topological product, and p_λ the projection from S to S_λ. Define a σ-algebra $\mathscr{B}$ on S by

   $$\mathscr{B} = \bigvee_\lambda p_\lambda^{-1}(\mathscr{B}(S_\lambda)).$$

Prove that $\mathscr{B}(S) \supset \mathscr{B}$ in general and that $\mathscr{B}(S) = \mathscr{B}$ if Λ is countable and every S_λ has a countable open base.

(iv) Suppose in (iii) that every S_λ contains at least two points and A is not countable. Prove that

$$\mathscr{B}(S) \neq \mathscr{B}.$$

[*Hint*: A set $B \subset S$ is called σ-*determined* if there exists a countable set $\Lambda_0 = \Lambda_0(B) \subset \Lambda$ such that

$$p_\lambda(s) = p_\lambda(s')(\lambda \in \Lambda_0) \Rightarrow 1_B(s) = 1_B(s').$$

The class $\mathscr{B}_0$ of all σ-determined sets is a σ-algebra and

$$\mathscr{B}_0 \supset p_\lambda^{-1}(\mathscr{B}(S_\lambda)), \qquad \lambda \in \Lambda,$$

and so

$$\mathscr{B}_0 \supset \mathscr{B}.$$

This implies that every element of $\mathscr{B}$ is σ-determined. It follows from our assumption that every singleton $\{s\}$ in S is not σ-determined. Hence $\{s\}$ does not belong to $\mathscr{B}$ but to $\mathscr{B}(S)$, because it is a closed set.]

(v) Prove that

$$\mathscr{B}_n \in \mathbf{B}(\Delta_n) \ (n = 1, 2, \dots) \Rightarrow \bigvee_n \mathscr{B}_n \in \mathbf{B}\left(\bigvee_n \Delta_n\right).$$

(vi) Prove that $\sigma[X]$ is equivalent to $\bar{\sigma}[X]$ for every topological random variable X.

(vii) Prove that if $\mathscr{B}_\lambda$ is equivalent to $\mathscr{B}'_\lambda$ for every λ, then $\bigvee_\lambda \mathscr{B}_\lambda$ is equivalent to $\bigvee_\lambda \mathscr{B}'_\lambda$. [*Hint*: $(\bigvee_\lambda \mathscr{B}_\lambda) \vee \mathbf{2} = \bigvee_\lambda (\mathscr{B}_\lambda \vee \mathbf{2}).$]

(viii) Prove that if $X(\omega) = (X_1(\omega), X_2(\omega), \dots)$, then P^X is the Lebesgue extension of its restriction to $\mathscr{B} \equiv \bigvee_n p_n^{-1}(\mathscr{D}(P^{X_n}))$. [*Hint*: Use Theorem 3.1.2, observing that

$$\sigma[\mathscr{A}] \subset \bigvee_n p_n^{-1}(\mathscr{D}(P^{X_n})) \subset \mathscr{D}(P^X)$$

for the $\mathscr{A}$ used in the definition of the joint variable $X(\omega)$ (Section 3.2).]

3.4   Independence

In this section a σ-subalgebra of $\mathcal{D}(P)$ is called simply a σ-algebra. A family of σ-algebras, $\{\mathcal{B}_\lambda, \lambda \in \Lambda\}$ is called *independent* if for every finite set $\{\lambda_1, \lambda_2, \ldots, \lambda_n\} \subset \Lambda$ we have

$$P\left(\bigcap_{i=1}^n B_i\right) = \prod_{i=1}^n P(B_i), \qquad \left(B_i \in \mathcal{B}_{\lambda_i}; i = 1, 2, \ldots, n\right).$$

Immediately from this definition we obtain the following:

Theorem 3.4.1.
 (i) A family of σ-algebras is independent if and only if every finite subfamily is independent.
 (ii) Every subfamily of an independent family of σ-algebras is independent.
 (iii) If $\mathcal{B}_\lambda$ is equivalent to $\mathcal{B}_\lambda$ for every $\lambda \in \Lambda$, then the independence of $\{\mathcal{B}_\lambda, \lambda \in \Lambda\}$ is equivalent to that of $\{\mathcal{B}'_\lambda, \lambda \in \Lambda\}$.
 (iv) If $\mathcal{B}'_\lambda \subset \mathcal{B}_\lambda$ for every $\lambda \in \Lambda$, then the independence of $\{\mathcal{B}_\lambda, \lambda \in \Lambda\}$ implies that of $\{\mathcal{B}'_\lambda, \lambda \in \Lambda\}$.

Theorem 3.4.2. Suppose that $\mathcal{A}_\lambda$ is a multiplicative class and $\mathcal{B}_\lambda = \sigma[\mathcal{A}_\lambda]$ for every $\lambda \in \Lambda$. If for every finite set $\{\lambda_1, \lambda_2, \ldots, \lambda_n\} \subset \Lambda$ we have

$$P\left(\bigcap_{i=1}^n A_i\right) = \prod_{i=1}^n P(A_i) \qquad \left(A_i \in \mathcal{A}_{\lambda_i}; i = 1, 2, \ldots, n\right),$$

then $\{\mathcal{B}_\lambda, \lambda \in \Lambda\}$ is independent.

Proof: It suffices to prove the following proposition for every m.
 (I_m) For every $n \geq m + 1$ it holds that

$$P\left(\bigcap_{i=1}^m B_i \cap \bigcap_{j=m+1}^n A_j\right) = \prod_{i=1}^m P(B_i) \prod_{j=m+1}^n P(A_j), \qquad B_i \in \mathcal{B}_{\lambda_i}, A_j \in \mathcal{A}_{\lambda_j}.$$

In case $m = 0$, the first intersection is Ω and the first product is 1 by convention. Similarly, for the case $m = n$.

 (I_0) is the assumption of the theorem. Assuming (I_m) to hold, we will prove that for $n \geq m + 2$ it holds that

$$P\left(\bigcap_{i=1}^{m+1} B_i \cap \bigcap_{j=m+2}^n A_j\right) = \prod_{i=1}^{m+1} P(B_i) \prod_{j=m+2}^n P(A_j),$$

$$B_i \in \mathcal{B}_{\lambda_i}, \quad A_j \in \mathcal{A}_{\lambda_j}.$$

Fix every B_i ($i \neq 1$) and every A_j and denote by $\mathscr{B}$ the class of all B_1 for which this equality holds. To complete the proof it remains only to prove that $\mathscr{B} \supset \mathscr{B}_{\lambda_1}$. $\mathscr{B}$ is closed under proper differences and countable disjoint unions by the σ-additivity of P. Since the equality holds for $B_1 \in \mathscr{A}_{\lambda_1}$ by (I_m), $\mathscr{B}$ is a Dynkin class including $\mathscr{A}_{\lambda_1}$. Hence $\mathscr{B} \supset \sigma[\mathscr{A}_{\lambda_1}] = \mathscr{B}_{\lambda_1}$ by the Dynkin class theorem. ∎

Theorem 3.4.3. Suppose that $\Lambda = \sum_{\mu \in M} \Lambda_\mu$ and $\mathscr{B}_\lambda$ is a σ-algebra for every $\lambda \in \Lambda$. We set

$$\mathscr{B}'_\mu = \bigvee_{\lambda \in \Lambda_\mu} \mathscr{B}_\lambda, \qquad \mu \in M.$$

Then $\{\mathscr{B}_\lambda, \lambda \in \Lambda\}$ is independent if and only if
(i) $\{\mathscr{B}_\lambda, \lambda \in \Lambda_\mu\}$ is independent for every $\mu \in M$, and
(ii) $\{\mathscr{B}'_\mu, \mu \in M\}$ is independent.

Proof: It suffices to prove the case where Λ is a finite set; then Λ_μ and M are also finite sets. Suppose that $\{\mathscr{B}_\lambda, \lambda \in \Lambda\}$ is independent. Being a subfamily of this family, $\{\mathscr{B}_\lambda, \lambda \in \Lambda_\mu\}$ is independent for every $\mu \in M$. Let $\mathscr{A}_\mu$ be the family of all sets A'_μ of the form

$$A'_\mu = \bigcap_{\lambda \in \Lambda_\mu} B_\lambda, \qquad B_\lambda \in \mathscr{B}_\lambda.$$

Then A'_μ is a multiplicative class generating $\mathscr{B}'_\mu$. By the independence of $\{\mathscr{B}_\lambda, \lambda \in \Lambda\}$ we obtain

$$P\left(\bigcap_{\lambda \in \Lambda} B_\lambda\right) = P\left(\bigcap_{\mu \in M} A'_\mu\right) = \prod_{\mu \in M} \prod_{\lambda \in \Lambda_\mu} P(B_\lambda) = \prod_{\lambda \in \Lambda} P(B_\lambda),$$

which implies the independence of $\{\mathscr{B}'_\mu, \mu \in M\}$ by the last theorem. Thus the "only if" part is proved. To prove the "if" part, take $B_\lambda \in \mathscr{B}_\lambda$ and define $A'_\mu \in \mathscr{B}'_\mu$ as above. Then

$$P(A'_\mu) = \prod_{\lambda \in \Lambda_\mu} P(B_\lambda), \qquad P\left(\bigcap_{\mu \in M} A'_\mu\right) = \prod_{\mu \in M} P(A'_\mu)$$

by the assumption and so

$$P\left(\bigcap_{\lambda \in \Lambda} B_\lambda\right) = P\left(\bigcap_{\mu \in M} A'_\mu\right) = \prod_{\mu \in M} \prod_{\lambda \in \Lambda_\mu} P(B_\lambda) = \prod_{\lambda \in \Lambda} P(B_\lambda),$$

which completes the proof of the "if" part. ∎

Theorem 3.4.4 (Kolmogorov's zero-one law). If $\{\mathscr{B}_1, \mathscr{B}_2, \ldots\}$ is independent, then

$$\limsup_{n \to \infty} B_n \left(= \bigwedge_k \bigvee_{n > k} \mathscr{B}_n \right) \subset 2;$$

that is, every set belonging to the left-hand side has probability 0 or 1.

Proof: Let

$$\mathscr{C}_k = \bigvee_{n > k} \mathscr{B}_n, \qquad \mathscr{C} = \bigwedge_k \mathscr{C}_k = \bigwedge_k \bigvee_{n > k} \mathscr{B}_n, \qquad \mathscr{B} = \bigvee_n \mathscr{B}_n.$$

Setting

$$\Lambda = \{1, 2, 3, \ldots\}, \qquad M = \{1, 2, \ldots, k+1\},$$

$$\Lambda_1 = \{1\}, \Lambda_2 = \{2\}, \ldots, \Lambda_k = \{k\}, \Lambda_{k+1} = \{k+1, k+2, \ldots\}$$

in the last theorem, we can see that $\{\mathscr{B}_1, \mathscr{B}_2, \ldots, \mathscr{B}_k, \mathscr{C}_k\}$ is independent. This implies that $\{\mathscr{B}_1, \mathscr{B}_2, \ldots, \mathscr{B}_k, \mathscr{C}\}$ is independent, because $\mathscr{C} \subset \mathscr{C}_k$. Since this holds for every k, $\{\mathscr{B}_1, \mathscr{B}_2, \ldots, \mathscr{C}\}$ is independent. Hence $\{\mathscr{B}, \mathscr{C}\}$ is independent by the last theorem. Since $\mathscr{C} \subset \mathscr{B}$, $\{\mathscr{C}, \mathscr{C}\}$ is independent. If $B \in \mathscr{C}$, then

$$P(B) = P(B \cap B) = P(B)P(B) \text{ (i.e., } P(B) = 0 \text{ or } 1). \qquad \blacksquare$$

A family of random variables $\{X_\lambda(\omega), \lambda \in \Lambda\}$ is called independent if $\{\bar{\sigma}[X_\lambda], \lambda \in \Lambda\}$ is independent.

Theorem 3.4.5.
   (i) If $\{X_\lambda(\omega), \lambda \in \Lambda\}$ is independent and if Y_λ is a daughter of X_λ for every λ, then $\{Y_\lambda(\omega), \lambda \in \Lambda\}$ is independent.
   (ii) $\{X_{\lambda n}(\omega), \lambda \in \Lambda, n = 1, 2, \ldots\}$ is independent if and only if (a) $\{X_{\lambda 1}(\omega), X_{\lambda 2}(\omega), \ldots\}$ is independent for every λ and (b) the family of the joint variable $X_\lambda(\omega) = (X_{\lambda 1}(\omega), X_{\lambda 2}(\omega), \ldots), \lambda \in \Lambda$, is independent.
   (iii) If $X(\omega)$ is the joint variable of an independent sequence of random variables $\{X_1(\omega), X_2(\omega), \ldots\}$, then P^X is the complete direct product of P^{X_n} $(n = 1, 2, \ldots)$.

Proof:
   (i) Note that if Y is a daughter of X, then $\bar{\sigma}[Y] \subset \bar{\sigma}[X]$ and use Theorem 3.4.1(iv).
   (ii) It suffices to check that if $X(\omega) = (X_1(\omega), X_2(\omega), \ldots)$ then $\bar{\sigma}[X]$ is equivalent to $\bigvee_n \bar{\sigma}[X_n]$; once this is done, part (ii) turns out to be a special case of Theorem 3.4.3. $\mathscr{D}(P^X)$ is equivalent to

$V_n p_n^{-1}(\mathscr{D}(P^{X_n}))$ with respect to P^X by Exercise 3.3(viii). Hence $\bar{\sigma}[X] = X^{-1}(\mathscr{D}(P^X))$ is equivalent to

$$X^{-1}\left(\bigvee_n p_n^{-1}(\mathscr{D}(P^{X_n}))\right) = \bigvee_n X_n^{-1}(\mathscr{D}(P^{X_n})) = \bigvee_n \bar{\sigma}[X_n]$$

with respect to P.

(iii) Let $B_n \in \mathscr{D}(P^{X_n})$ $(n = 1, 2, \ldots)$. Then

$$P^X(B_1 \times B_2 \times \cdots) = P\left(X^{-1}(B_1 \times B_2 \times \cdots)\right)$$

$$= P\left(\bigcap_n X_n^{-1}(B_n)\right)$$

$$= \prod_n P\left(X_n^{-1}(B_n)\right) = \prod_n P^{X_n}(B_n).$$

This implies that the direct product measure $\prod_n P^{X_n}$ is the restriction of P^X to $V_n p_n^{-1}(\mathscr{D}(P^{X_n}))$, whose Lebesgue extension is P^X by Exercise 3.3(viii). ∎

A family of topological random variables $\{X_\lambda(\omega), \lambda \in \Lambda\}$ is independent if and only if $\{\sigma[X_\lambda] \lambda \in \Lambda\}$ is independent, as we can see from the equivalence of $\bar{\sigma}[X_\lambda]$ and $\sigma[X_\lambda]$ and Theorem 3.4.1(iii). Also Theorem 3.4.2 implies that $\{X_\lambda(\omega), \lambda \in \Lambda\}$ is independent if

$$P\left(\bigcap_{i=1}^n X_{\lambda_i}^{-1}(G_i)\right) = \prod_{i=1}^n P\left(X_{\lambda_i}^{-1}(G_i)\right)$$

whenever G_i is an open subset of the range space of $X_{\lambda_i}(\omega)$ for every i.

A family of P-measurable sets $\{A_\lambda, \lambda \in \Lambda\}$ is called independent if $\{1_{A_\lambda}(\omega), \lambda \in \Lambda\}$ is independent; or equivalently if $\{\mathscr{B}_\lambda = \{\varnothing, A_\lambda, A_\lambda^c, \Omega\}, \lambda \in \Lambda\}$ is independent. A finite family of P-measurable sets $\{A_1, A_2, \ldots, A_n\}$ is independent if and only if

$$P\left(\bigcap_{\nu=1}^k A_{i_\nu}\right) = \prod_{\nu=1}^k P\left(A_{i_\nu}\right)$$

holds for every $k = 2, 3, \ldots, n$ and every $1 \le i_1 < i_2 < \cdots < i_k \le n$, as is clear by Theorem 3.4.2(i). (See Theorem 1.5.6.) The following theorem is a generalization of Theorem 1.6.1.

Theorem 3.4.6 (Multiplicativity of mean values). Suppose that $\{X_1(\omega), X_2(\omega), \ldots, X_n(\omega)\}$ is an independent family of real (or complex) random variables. Then

$$E\left(\prod_k X_k\right) = \prod_k EX_k$$

holds in each one of the following cases:

(i) $X_k(\omega) \geq 0$ $(k = 1, 2, \ldots, n)$;

(ii) $E|X_k| < \infty$ $(k = 1, 2, \ldots, n)$.

Proof: In case X_k is the indicator of a P-measurable set A_k for every k, the equality follows from

$$EX_k = P(A_k), \qquad E\left(\prod_k X_k\right) = P\left(\bigcap_k A_k\right).$$

Hence we can prove the equality for the case (i) by using the properties of mean values (Theorem 3.2.2). To prove the equality for case (ii), use the decompositions:

$$X_k = X_k^+ - X_k^-, \qquad \text{if } X_k \text{ is real}$$

$$X_k = \mathrm{Re}(X_k) + i\,\mathrm{Im}(X_k) \qquad \text{if } X_k \text{ is complex.} \qquad \blacksquare$$

The following theorem claims that independence is inherited by limit variables.

Theorem 3.4.7. Suppose that for every $\lambda \in \Lambda$, $X_\lambda^{(n)}(\omega)$ and $X_\lambda(\omega)$ are topological random variables taking values in a metric space S_λ. If $\{X_\lambda^{(n)}(\omega), \lambda \in \Lambda\}$ is independent for every n and if $X_\lambda^{(n)}(\omega) \to X_\lambda(\omega)$ a.s. for every λ, then $\{X_\lambda(\omega), \lambda \in \Lambda\}$ is independent.

Proof: It suffices to discuss the case where Λ is a finite set. Let f_λ: $S_\lambda \to \mathbb{R}^1$ be bounded and continuous. Then

$$\left\{Y_\lambda^{(n)}(\omega) = f_\lambda(X_\lambda^{(n)}(\omega)), \lambda \in \Lambda\right\}$$

is independent by Theorem 3.4.5(i). Hence the last theorem ensures that

$$E\left(\prod_\lambda Y_\lambda^{(n)}\right) = \prod_\lambda E\left(Y_\lambda^{(n)}\right).$$

Letting $n \to \infty$ and using the dominated convergence theorem (Theorem 3.2.2(vii)) we obtain

$$E\left(\prod_\lambda f_\lambda(X_\lambda(\omega))\right) = \prod_\lambda E(f_\lambda(X_\lambda(\omega))).$$

Since the indicator of an open subset of a metric space is represented as the limit of continuous functions with values between 0 and 1, we can use the dominated convergence theorem to conclude that the equality above holds for $f_\lambda = 1_{G_\lambda}$ (G_λ: open in S_λ); that is:

$$P\left(\bigcap_\lambda X_\lambda^{-1}(G_\lambda)\right) = \prod_\lambda P(X_\lambda^{-1}(G_\lambda)).$$

Hence $\{X_\lambda(\omega), \lambda \in \Lambda\}$ is independent. ∎

A family of decompositions $\{\Delta_\lambda, \lambda \in \Lambda\}$ is called *independent* if the corresponding family of σ-algebras $\{\mathscr{B}_{\Delta_\lambda}, \lambda \in \Lambda\}$ is independent. It is obvious that

$$\{\Delta_\lambda, \lambda \in \Lambda\} \perp\!\!\!\perp \Leftrightarrow \{X_{\Delta_\lambda}, \lambda \in \Lambda\} \perp\!\!\!\perp,$$

$$\{X_\lambda, \lambda \in \Lambda\} \perp\!\!\!\perp \Leftrightarrow \{\Delta_{X_\lambda}, \lambda \in \Lambda\} \perp\!\!\!\perp,$$

where $\perp\!\!\!\perp$ means independence.

A family of events $\{\alpha_\lambda(\omega), \lambda \in \Lambda\}$ is called *independent* if the corresponding family of P-measurable sets $\{\{\alpha_\lambda\}, \lambda \in \Lambda\}$ is independent. The Borel–Cantelli lemma is strengthened in the following form for an independent sequence of events.

Theorem 3.4.8 (Borel's theorem). If $\{\alpha_1, \alpha_2, \ldots\}$ is independent, then

$$\sum_n P\{\alpha_n\} < \infty \Rightarrow P\{\alpha_n \text{ i.o.}\} = 0, \ P\{\alpha_n^- \text{ f.e.}\} = 1,$$

$$\sum_n P\{\alpha_n\} = \infty \Rightarrow P\{\alpha_n \text{ i.o.}\} = 1, \ P\{\alpha_n^- \text{ f.e.}\} = 0.$$

Proof: The first assertion is obvious by the Borel–Cantelli lemma. We will prove the second assertion. Letting $A_n = \{\alpha_n\}$, we obtain

$$P\{\alpha_n^- \text{ f.e.}\} = P\left\{\bigcup_k \bigcap_{n \geq k} A_n^c\right\} \leq \sum_k P\left(\bigcap_{n \geq k} A_n^c\right),$$

$$P\left(\bigcap_{n \geq k} A_n^c\right) \leq P\left(\bigcap_{n=k}^m A_n^c\right) = \prod_{n=k}^m P(A_n^c)$$

$$= \prod_{n=k}^m (1 - P(A_n)).$$

The last product tends to 0 by the assumption $\sum P(A_n) = \infty$. Hence $P\{\alpha_n^- \text{ f.e.}\} = 0$; that is, $P\{\alpha_n \text{ i.o.}\} = 1$. ∎

Remark: When a family of σ-algebras $\{B_1, B_2, \ldots\}$ is independent, we say that $B_1, B_2, \ldots$ are independent. If B_1, B_2 are independent, B_1 is said to be independent of B_2. Similarly for random variables, decompositions, and events.

Exercise 3.4.

(i) Prove that if $\mathscr{B}_k$ is independent of $\mathscr{C}_k = \bigvee_{n>k} \mathscr{B}_n$ for every k, then $\mathscr{B}_1, \mathscr{B}_2, \ldots$ are independent. [*Hint*: Noting that $\mathscr{C}_1 = \mathscr{B}_2 \vee \mathscr{C}_2$ and using Theorem 3.4.3 we can deduce the independence of $\{\mathscr{B}_1, \mathscr{B}_2, \mathscr{C}_2\}$ from the assumption. Repeat this argument to verify the independence of $\{\mathscr{B}_1, \mathscr{B}_2, \ldots, \mathscr{B}_k, \mathscr{C}_k\}$.]

(ii) Prove that if $X_1, X_2, \ldots, X_n$ are independent real random variables, then $Y_1 = X_1 + X_2 + \cdots + X_k$, $Y_2 = X_{k+1} + X_{k+2} + \cdots + X_n$ are independent. [*Hint*: Use Theorem 3.4.3, observing that

$$\sigma[Y_1] \subset \sigma[(X_1, \ldots, X_k)] = \bigvee_{i=1}^{n} \sigma[X_i] \qquad \text{and}$$

$$\sigma[Y_2] \subset \bigvee_{j=k+1}^{n} \sigma[X_j].]$$

(iii) Prove that if a real random variable $X(\omega)$ is measurable **2**, then X is a constant a.s. [*Hint*: Since $X^{-1}(-\infty, x] \in \mathbf{2}$, $P\{X \leq x\} = 0$ or 1. Since $P\{X \leq x\}$ is increasing in x, there exists $a \in \mathbb{R}^1$ such that $P\{X \leq a - \varepsilon\} = 0$ and $P\{X \leq a + \varepsilon\} = 1$ ($\varepsilon > 0$). Hence $P(X = a) = 1$.]

(iv) Let $\{X_n(\omega), n = 1, 2, \ldots\}$ be an independent sequence of real random variables. Prove that the following events have probability 0 or 1.
(a) $X_n(\omega)$ is convergent as $n \to \infty$.
(b) $X_n(\omega)$ is bounded. (The bound may depend on ω.)
(c) $\sum_n |X_n(\omega)| < \infty$.
(d) $n^{-1} \sum_{k=1}^{n} X_k(\omega) \to 0$ ($n \to \infty$).
(e) $X_n(\omega) > 0$ i.o..
(f) $X_n(\omega) > n$ f.e..
[*Hint*: Let $\mathscr{B}_n = \sigma[X_n]$. Observe that every event belongs to $\limsup \mathscr{B}_n$ and use the zero-one law.]

(v) Let $X_1(\omega), X_2(\omega), \ldots$ be independent real random variables. Prove that

$$Y(\omega) = \limsup_{n \to \infty} X_n(\omega)$$

is a constant a.s. [*Hint*:

$$\{Y(\omega) \leq y\} \in \limsup_{n \to \infty} \mathscr{B}_n \subset \mathbf{2}, \qquad \mathscr{B}_n = \sigma[X_n].]$$

(vi) Give an example of a decreasing sequence of separable decompositions whose greatest lower bound is not separable. [*Hint*: Let $\Omega = \mathbb{R}^{\infty}$, $P = \prod \mu_n$ and $X_n(\omega)$ be the nth component of ω. We assume that every μ_n is a continuous distribution on $\mathbb{R}^1$. Then $X_1(\omega), X_2(\omega), \ldots$ are independent real random variables on (Ω, P) such that

$$P\{ X_n = x_n \} = 0, \qquad n = 1, 2, \ldots; x_n \in \mathbb{R}.$$

Let Δ_n be a decomposition under the joint variable $(X_n, X_{n+1}, \ldots)$. Then $\{\Delta_n\}$ is a decreasing sequence of separable decompositions. But $\Delta = \wedge_n \Delta_n$ is not separable. Suppose to the contrary that Δ is separable. Then it corresponds to a real random variable, say, X. For every n, $\Delta \prec \Delta_n$ and so X is measurable $\vee_{k \geq n} \sigma[X_k]$. This implies that X is measurable **2** (Theorem 3.4.4). Hence there exists a real number x such that $P\{ X = x \} = 1$. Since $\{ X = x \} \in \Delta$, we can find $A_n \in \Delta_n$, $n = 1, 2, \ldots$ such that

$$\{ X = x \} = \bigcup_n A_n \qquad \text{(Theorem 3.3.2(iii))}.$$

Since $\Delta_n \succ \Delta_{X_n}$, $A_n \subset \{ X_n = x_n \}$ $(x_n \in \mathbb{R})$, and so $P(A_n) = 0$. Hence we obtain

$$1 = P\{ X = x \} \leq \sum_n P(A_n) = 0,$$

which is a contradiction.]

3.5 Conditional probability measures

Let (Ω, P) be the base probability space and

$$\Delta = \{ \xi_1, \xi_2, \ldots, \xi_n \}$$

be a finite decomposition of Ω. To make it easier for the reader to understand the essential points we assume that

$$P(\xi_i) > 0, \qquad i = 1, 2, \ldots, n.$$

For each ξ_i we define a probability measure P_{ξ_i} by

$$(C_1) \qquad P_{\xi_i}(A) = \frac{P(A)}{P(\xi_i)} \qquad (A \subset \xi_i, A \in \mathscr{D}(P)),$$

and call $P_{\xi_i}(A)$ the probability of A under the condition that a sample point in ξ_i is realized or simply the conditional probability of A under ξ_i. Also, the probability measure P_{ξ_i} is called the conditional probability measure under ξ_i.

Let P^Δ be the probability measure on Δ induced from P in the natural way; that is,

$$P^\Delta(E) = P\left(\sum_{\xi \in E} \xi\right) = \sum_{\xi \in E} P(\xi), \quad E \subset \Delta.$$

Let X_Δ be the random variable corresponding to Δ; that is,

$$X_\Delta(\omega) = \text{the element of } \Delta \text{ containing } \omega.$$

Then

$$P^\Delta = PX_\Delta^{-1} = P^{X_\Delta}.$$

Thus we have derived P^Δ and $\{P_\xi, \xi \in \Delta\}$ from P. Conversely we can express P in terms of P^Δ and P_ξ as follows:

$$P(A) = \sum_{i=1}^{n} P_{\xi_i}(A \cap \xi_i) P^\Delta(\xi_i), \quad A \in \mathscr{D}(P)$$

or in the integral form

$$(\text{C}_2) \quad P(A) = \int_\Delta P_\xi(A \cap \xi) P^\Delta(d\xi), \quad A \in \mathscr{D}(P).$$

Suppose that for every ξ there exists a probability measure P_ξ for which (C_2) holds. If $A \subset \xi_i$, then $A \cap \xi = \emptyset$ for $\xi \neq \xi_i$, and so we obtain (C_1). Hence (C_1) and (C_2) are equivalent to each other for every finite decomposition Δ. Since (C_2) is meaningful for a general decomposition, it sounds reasonable to define the conditional probability measure P_ξ by (C_2).

Theorem 3.5. Let Δ be a separable decomposition of Ω. Then there exists a family $\{P_\xi, \xi \in \Delta\}$ such that (1) for every ξ, P_ξ is a separable perfect probability measure on Ω with $P_\xi(\xi) = 1$, (2) for every $A \in \mathscr{D}(P)$, $A \in \mathscr{D}(P_\xi)$ except on a P^Δ-null set in Δ (which may depend on A in general), (3) for every $A \in \mathscr{D}(P)$, $P_\xi(A)$ is P^Δ-measurable in ξ, and (4) condition (C_2) holds.

Such a family $\{P_\xi, \xi \in \Delta\}$ is essentially unique in the sense that if $\{P'_\xi, \xi \in \Delta\}$ is any other family satisfying these properties, then $P_\xi = P'_\xi$ a.e. (P^Δ) on Δ (or a.s. on (Δ, P^Δ) in the probability language).

Remark: $P_\xi(A)$ is defined on a set $\Delta' = \Delta'(A)$ with $P^\Delta(\Delta') = 1$ by (2). Hence condition (3) means that $P_\xi(A)$ coincides with a P^Δ-measurable function of $\xi \in \Delta$ on Δ', as is usual in measure theory.

Definition 3.5. The family $\{P_\xi, \ \xi \in \Delta\}$ introduced above is called the *conditional probability measures under the decomposition* Δ. Since P_ξ depends on Δ, we often denote it by $P_{\Delta, \xi}$.

Remark: In view of the fact $P_\xi(\xi) = 1$, P_ξ may be regarded as a separable perfect measure on ξ by restricting it to ξ.

Proof of the theorem: Being separable and perfect, P is strictly isomorphic to a K-regular probability measure on a subset of $\mathbb{R}^1$ (Theorem 3.1.1.). Hence we can assume without any loss of generality that

$$\Omega \subset \mathbb{R}^1, \qquad P: K\text{-regular probability measure on } \Omega.$$

Since Δ is separable, the corresponding X_Δ introduced above is a Δ-valued random variable and will be denoted simply by X hereafter. As we have mentioned above,

$$P^\Delta = PX^{-1} = P^X.$$

Hence P^Δ is a separable perfect probability measure on Δ.

First we will prove the uniqueness part of the theorem. Suppose that P_ξ and P'_ξ satisfy the conditions in the theorem. Replacing A by $A \cap X^{-1}(E)$ $(E \in \mathscr{D}(P^\Delta))$ we obtain

$$P(A \cap X^{-1}(E)) = \int_\Delta P_\xi(A \cap X^{-1}(E) \cap \xi) P^\Delta(d\xi)$$

$$= \int_\Delta P'_\xi(A \cap X^{-1}(E) \cap \xi) P^\Delta(d\xi).$$

According to whether $\xi \in E$ or not, we have

$$\xi \subset X^{-1}(E) \qquad \text{or} \qquad \xi \cap X^{-1}(E) = \varnothing$$

and so

$$\int_E P_\xi(A \cap \xi) P^\Delta(d\xi) = \int_E P'_\xi(A \cap \xi) P^\Delta(d\xi), \qquad E \in \mathscr{D}(P^\Delta).$$

This implies that

$$P_\xi(A \cap \xi) = P'_\xi(A \cap \xi) \quad \text{a.e.}(P^\Delta).$$

Since the exceptional ξ-set for this equality depends on A, we denote it by $N(A)$. $N(A)$ is a P^Δ-null set. Since Ω is a subset of $\mathbb{R}^1$, the Borel system $\mathscr{B}(\Omega)$ is generated by the class

$$\Omega_r = \Omega \cap (-\infty, r], \qquad r \in \mathbb{Q}.$$

The union $N = \bigcup_r N(\Omega_r)$ is also a P^Δ-null set, and

$$P_\xi(\Omega_r \cap \xi) = P'_\xi(\Omega_r \cap \xi), \qquad r \in \mathbf{Q}$$

holds for every $\xi \in \Delta - N$. Since the Borel system $\mathscr{B}(\xi)$ is generated by the class $\Omega_r \cap \xi$, $r \in \mathbf{Q}$, the Dynkin class theorem ensures that

$$P_\xi(A) = P'_\xi(A), \qquad A \in \mathscr{B}(\xi),$$

Being separable and perfect, both P_ξ and P'_ξ are regular. Hence we obtain

$$P_\xi = P'_\xi, \qquad \xi \in \Delta - N.$$

Next we will prove the existence part. If there exists a family $\{ P_\xi, \xi \in \Delta \}$ satisfying the conditions in the theorem, it must satisfy the equalities

$$P(A \cap X^{-1}(E)) = \int_E P_\xi(A \cap \xi) P^\Delta(d\xi), \qquad A \in \mathscr{D}(P); E \in \mathscr{D}(P^X),$$

as we have observed above. Aiming at these equalities, we will determine P_ξ. Fixing A for the moment, we set

$$Q_A(E) = P(A \cap X^{-1}(E)), \qquad E \in \mathscr{D}(P^\Delta).$$

Then Q_A is a finite measure on Δ and

$$\mathscr{D}(Q_A) - \mathscr{D}(P^\Delta),$$
$$P^\Delta(E) = 0 \Rightarrow Q_A(E) \leqq P(X^{-1}(E)) = P^\Delta(E) = 0.$$

The well-known Randon–Nikodym theorem in measure theory ensures the existence of a P^Δ-measurable function $f_A(\xi)$ satisfying

$$(\mathrm{D}_1) \qquad Q_A(E)\big(= P(A \cap X^{-1}(E))\big) = \int_E f_A(\xi) P^\Delta(d\xi), \qquad E \in \mathscr{D}(P^\Delta).$$

This function $f_A(\xi)$ is regarded as a candidate for $P_\xi(A \cap \xi)$. For each set A there are many such functions, any two of which are equal to each other a.e. (P^Δ). As is shown below, we can choose any one of these functions $f_A(\xi)$ and modify it on a P^Δ-null set for each A so that $f_A(\xi)$ can be written in the form $P_\xi(A \cap \xi)$. It is obvious that such a modification preserves the relation (D_1).

Observing that $0 \leq Q_A(E) \leq P^\Delta(E)$, we obtain

$$0 \leqq f_A(\xi) \leqq 1 \text{ a.e. } (P^\Delta).$$

Putting $A = \phi$ or Ω, and $E = \Delta$ in (D_1), we obtain

$$f_\varnothing(\xi) = 0, \qquad f_\Omega(\xi) = 1 \text{ a.e.} (P^\Delta).$$

We modify $f_A(\xi)$ so that these conditions are satisfied for every $\xi \in \Delta$.

Putting $A = \Omega_r \equiv \Omega \cap (-\infty, r]$ and denoting $f_{\Omega_r}(\xi)$ by $F_\xi(r)$ we obtain

$(D_2) \quad P\big(\Omega_r \cap X^{-1}(E)\big) = \int_E F_\xi(r) P^\Delta(d\xi), \quad E \in \mathscr{D}(P^\Delta).$

Hence for each pair $r < r'$ we have $F_\xi(r) \leq F_\xi(r')$ a.e.(P^Δ). Since $\mathbf{Q} \times \mathbf{Q}$ is countable, we can assume that these inequalities hold for every pair $r < r'$ and every ξ. Then

$$F_\xi(r+) = \lim_{n \to \infty} F_\xi(r + n^{-1}), \quad \text{and} \quad F_\xi(\pm\infty) = \lim_{n \to \infty} F_\xi(\pm n)$$

exist, and it holds that

$$F_\xi(r+) \geq F_\xi(r), \quad 0 \leq F_\xi(-\infty) \leq F_\xi(\infty) \leq 1.$$

Observing that $\Omega_{r+n^{-1}} \downarrow \Omega_r$, we obtain

$$\int_\Delta F_\xi(r+) P^\Delta(d\xi) = \lim_{n \to \infty} \int_\Delta F_\xi(r + n^{-1}) P^\Delta(d\xi)$$

$$= \lim_{n \to \infty} P(\Omega_{r+n^{-1}}) = P(\Omega_r)$$

$$= \int_\Delta F_\xi(r) P^\Delta(d\xi)$$

and so $F_\xi(r+) = F_\xi(r)$ a.e. (P^Δ) for every r. Since $\mathbf{Q}$ is countable, we can assume that this holds for every r and every ξ. Since $\Omega_{-n} \downarrow \varnothing$ and $\Omega_n \uparrow \Omega$ $(n \to \infty)$, we can assume that

$$F_\xi(-\infty) = 0 \quad \text{and} \quad F_\xi(\infty) = 1$$

for every ξ. Hence there exist a regular probability measure μ_ξ on $\mathbb{R}^1$ such that

$$F_\xi(r) = \mu_\xi(-\infty, r], \quad r \in \mathbf{Q}.$$

Recalling the definition of $F_\xi(r)$, we obtain

$$P\big(\Omega \cap (-\infty, r] \cap X^{-1}(E)\big) = \int_E \mu_\xi(-\infty, r] P^\Delta(d\xi), \quad r \in \mathbf{Q}.$$

This implies

$(D_3) \quad P\big(\Omega \cap A \cap X^{-1}(E)\big) = \int_E \mu_\xi(A) P^\Delta(d\xi), \quad A \in \mathscr{B}^1$

by the Dynkin class theorem.

The μ_ξ thus obtained is a probability measure on $\mathbb{R}^1$. Now we will prove that μ_ξ is concentrated on Ω. Since P is a K-regular probability measure on Ω, we can find an increasing sequence of compact sets

$K_n \subset \Omega$ $(n = 1, 2, \dots)$ such that

$$\lim_{n \to \infty} P(K_n) = P(\Omega) = 1.$$

Then

$$\int_{\Delta} \lim_{n \to \infty} \mu_{\xi}(K_n) P^{\Delta}(d\xi) = \lim_{n \to \infty} \int_{\Delta} \mu_{\xi}(K_n) P^{\Delta}(d\xi) = \lim_{n \to \infty} P(K_n) = 1,$$

and so

$$\lim_{n \to \infty} \mu_{\xi}(K_n) = 1 \text{ a.e. } (P^{\Delta}).$$

Since μ_{ξ} is complete, we obtain $\Omega \in \mathcal{D}(\mu_{\xi})$ and $\mu_{\xi}(\Omega) = 1$ a.e. (P^{Δ}). Hence we can assume that this holds for every ξ.

Let ν_{ξ} denote the restriction of μ_{ξ} to Ω. The domain $\mathcal{D}(\nu_{\xi})$ may depend on ξ but it is obvious that $\mathcal{D}(\nu_{\xi}) \supset \mathcal{B}(\Omega)$. Since every $B \in \mathcal{B}(\Omega)$ is written in the form

$$B = B' \cap \Omega, \qquad B' \in \mathcal{B}^1,$$

we obtain

$$\int_{E} \nu_{\xi}(B) P^{\Delta}(d\xi) = \int_{E} \mu_{\xi}(B') P^{\Delta}(d\xi) = P(\Omega \cap B' \cap X^{-1}(E))$$

$$= P(B \cap X^{-1}(E))$$

for every $E \in \mathcal{D}(P^{\Delta})$ and every $B \in \mathcal{B}(\Omega)$.

As we have mentioned above, $\mathcal{D}(\nu_{\xi}) \supset \mathcal{B}(\Omega)$. But it may not hold that $\mathcal{D}(\nu_{\xi}) \supset \mathcal{D}(P)$. However, we can prove that every $A \in \mathcal{D}(P)$ belongs to $\mathcal{D}(\nu_{\xi})$ except on a P^{Δ}-null ξ-set that may depend on A and that the following equality holds:

$$(\text{D}_4) \qquad \int_{E} \nu_{\xi}(A) P^{\Delta}(d\xi) = P(A \cap X^{-1}(E)).$$

Since P is a K-regular probability measure, we can find $B_1, B_2 \in B(\Omega)$ such that

$$B_1 \subset A \subset B_2, \qquad P(B_2 - B_1) = 0.$$

Since $B_1, B_2 \in \mathcal{B}(\Omega) \subset \mathcal{D}(\nu_{\xi})$, $\nu_{\xi}(B_2 - B_1)$ is defined and

$$\int_{\Delta} \nu_{\xi}(B_2 - B_1) P^{\Delta}(d\xi) = P(B_2 - B_1) = 0.$$

Hence $\nu_{\xi}(B_2 - B_1) = 0$ a.e. (P^{Δ}) and so

$$A \in \mathcal{D}(\nu_{\xi}), \qquad \nu_{\xi}(A) = \nu_{\xi}(B_1) = \nu_{\xi}(B_2) \text{ a.e. } (P^{\Delta}).$$

Therefore,

$$\int_E \nu_\xi(A) P^\Delta(d\xi) = \int_E \nu_\xi(B_1) P^\Delta(d\xi) = P(B_1 \cap X^{-1}(E))$$

$$= P(A \cap X^{-1}(E)).$$

This proves (D_4).

If we prove

$$\xi \in \mathscr{D}(\nu_\xi), \qquad \nu_\xi(\xi) = 1 \text{ a.e. } (P^\Delta),$$

then $\{ P_\xi \equiv \nu_\xi, \xi \in \Delta \}$ satisfies the conditions (1), (2), (3), and (4). Being separable, Δ is generated by a countable family:

$$A_n = X^{-1}(E_n), \qquad E_n \in \mathscr{D}(P^\Delta).$$

Setting $A = A_n$, $E = E_n$ in (D_4) we obtain

$$\int_{E_n} \nu_\xi(A_n) P^\Delta(d\xi) = P(A_n \cap X^{-1}(E_n)) = P(X^{-1}(E_n)) = P^\Delta(E_n)$$

and so

$$A_n \in \mathscr{D}(\nu_\xi), \qquad \nu_\xi(A_n) = 1 \text{ a.e. } (P^\Delta) \text{ on } E_n.$$

Similarly, we have

$$A_n^c \in \mathscr{D}(\nu_\xi), \qquad \nu_\xi(A_n^c) = 1 \text{ a.e. } (P^\Delta) \text{ on } E_n^c.$$

Denoting by N_n, N_n' the exceptional ξ-sets in the statements above, and letting

$$N = \bigcup_n (N_n \cup N_n') \qquad (P^\Delta(N) = 0),$$

we obtain

$$\xi \in E_n \Rightarrow A_n \in \mathscr{D}(\nu_\xi), \qquad \nu_\xi(A_n) = 1,$$

$$\xi \in E_n^c \Rightarrow A_n^c \in \mathscr{D}(\nu_\xi), \qquad \nu_\xi(A_n^c) = 1$$

for every $\xi \in \Delta - N$. Since $\{A_n\}$ generates Δ, every $\xi \in \Delta$ is represented in the form

$$\xi = \bigcap_n A_n'$$

where $A_n' = A_n$ or A_n^c according to whether $\xi \in E_n$ or E_n^c. From the fact just proved above we have

$$\nu_\xi(A_n') = 1, \qquad n = 1, 2, \ldots,$$

and so

$$\xi \in \mathscr{D}(\nu_\xi), \qquad \nu_\xi(\xi) = 1$$

for every $\xi \in \Delta - N$. This is true also for $\xi \in N$ by defining ν_ξ to be the δ-measure concentrated at a single point $\omega_0 = \omega_0(\xi)$ in ξ. ∎

3.5a   The conditional probability measure under a random variable

Let $X(\omega)$ be an S-valued random variable. Then

$$\Delta_X = \{ X^{-1}(s), s \in X(\Omega) \}$$

is a separable decomposition of Ω and we can define, for every s, the conditional probability measure $P_{\Delta_X, X^{-1}(s)}$, which is called the conditional probability measure under $X = s$, written $P_{X=s}$, in view of the fact that $X^{-1}(s) = \{ \omega | X(\omega) = s \}$. $P_{X=s}$ is a separable perfect measure on Ω concentrated on $X^{-1}(s)$ for every s. It is measurable $\bar{\sigma}(X)$ in s and has the property:

$$P(A \cap X^{-1}(E)) = \int_E P_{X=s}(A) P^X(ds),$$

because $P_{X=s}(A \cap X^{-1}(E)) = P_{X=s}(A)$ or 0 according to whether $s \in E$ or not. In the special case where $P\{ X(\omega) = s \} > 0$ we obtain

$$P_{X=s}(A) = \frac{P(\{ X=s \} \cap A)}{P\{ X=s \}}, \qquad A \in \mathscr{D}(P).$$

3.5b   The conditional probability measure under a σ-algebra

First we consider the case where $\mathscr{B}$ is determined by a separable decomposition Δ; that is, $\mathscr{B} = \mathscr{B}_\Delta$, or slightly more generally, $\mathscr{B}$ is subordinate to Δ; that is, $\Delta \subset \mathscr{B} \subset \mathscr{B}_\Delta$. In this case we define the conditional probability measure under $\mathscr{B}$, written $P_{\mathscr{B}, \omega}$, to be equal to $P_{\Delta, X_\Delta(\omega)}$, where $X_\Delta(\omega)$ is the element of Δ containing ω. Then $P_{\mathscr{B}, \omega}$ is a separable perfect probability measure concentrated on $X_\Delta(\omega) \in \Delta$. It is measurable $\mathscr{B}$ and has the property:

$$P(A \cap B) = E(P_{\mathscr{B}, \omega}(A), B), \qquad B \in \mathscr{B},$$

because $P_{\mathscr{B}, \omega}(A \cap B) = P_{\mathscr{B}, \omega}(A)$ or 0 according to whether $\omega \in B$ (i.e., $X_\Delta(\omega) \subset B$) or not.

Let $\mathscr{B}$ be a general σ-algebra consisting of P-measurable sets. Then there exists a sequence $\mathscr{A} = \{ A_n \}$ such that every set $B \in \mathscr{B}$ differs from a set $A \in \sigma(\mathscr{A})$ by a P-null set. To prove this we observe that $\mathscr{D}(P)$ is a separable metric space with metric

$$d(C, D) = P(C \Delta D) \qquad (\Delta: \text{symmetric difference}).$$

Hence we can find a countable class $\mathscr{A} = \{A_n\}$ such that every set $B \in \mathscr{B}$ is the d-limit of a subsequence $\{A_{n_k}\}$ of $\mathscr{A}$. Then $A = \limsup_k A_{n_k} \in \sigma[A]$ and B differs from A by a P-null set. Let Δ be the decomposition generated by $\{A_n\}$. We define

$$P_{\mathscr{B}, \omega} = P_{\mathscr{B}_\Delta, \omega}.$$

This definition is independent of the choice of $\mathscr{A}$ a.s., as we can see below.

Suppose that $\mathscr{A}' = \{A_n'\}$ is another sequence with the same property as $\mathscr{A}$, and determines a decomposition Δ'. Then for every n we can find $\mathscr{C} \equiv \{C_n\} \subset \sigma(\mathscr{A})$ and $\mathscr{C}' \equiv \{C_n'\} \subset \sigma[\mathscr{A}']$ such that

$$P(A_n' \Delta C_n) = 0, \qquad P(A_n \Delta C_n') = 0.$$

Let $\mathscr{D} \equiv \{D_k\} = \mathscr{A} \cup \mathscr{C}$ and $\mathscr{D}' \equiv \{D_n'\} = \mathscr{A}' \cup \mathscr{C}'$. Since

$$\mathscr{A} \subset \mathscr{D} \subset \sigma(\mathscr{A}), \qquad \mathscr{A}' \subset \mathscr{D}' \subset \sigma(\mathscr{A}'),$$

$\sigma(\mathscr{A})$ and $\sigma(\mathscr{A}')$ are generated by $\mathscr{D}$ and $\mathscr{D}'$. Hence Δ and Δ' are also determined by $\mathscr{D}$ and $\mathscr{D}'$, respectively. The ω-set $\Omega_1 = (\bigcup_n ((A_n' \Delta C_n) \cup (A_n \Delta C_n')))^c$ has P-measure 1. Then

$$\Delta \cap \Omega_1 = \Delta' \cap \Omega_1.$$

In other words, $\{\xi \cap \Omega_1 | \xi \in \Delta\} = \{\xi' \cap \Omega_1 | \xi' \in \Delta'\}$, and so

$$\mathscr{B}_\Delta \cap \Omega_1 = \mathscr{B}_{\Delta'} \cap \Omega_1.$$

Therefore $P_{B_\Delta, \omega}$ does not change by replacing Δ by Δ' whenever $\omega \in \Omega_1$.

The conditional probability measure $P_{\mathscr{B}, \omega}$ for a general σ-algebra $\mathscr{B}$ is a separable perfect probability measure on Ω. It is measurable $\mathscr{B}$ in ω and has the property:

$$P(A \cap B) = E(P_{\mathscr{B}, \omega}(A), B), \qquad B \in \mathscr{B}.$$

Contrary to the case where $\mathscr{B} = \mathscr{B}_\Delta$ we have no property of concentration in the strict sense mentioned above. The only result in this connection is as follows:

$$B \in \mathscr{B} \Rightarrow P_{\mathscr{B}, \omega}(B) = 1_B(\omega) \quad \text{a.s.}$$

This is immediate from the observation that

$$P(B) = P(B \cap B) = E(P_{\mathscr{B}}(B), B),$$
$$0 = P(B \cap B^c) = E(P_{\mathscr{B}}(B), B^c).$$

It should be noted that in the special case $\mathcal{B} = \mathcal{B}_\Delta$ or $\mathcal{B} = \bar{\sigma}(X)$ we have $P_{\Delta, \xi} = P_{\mathcal{B}, \omega}$, $\omega \in \xi$, or $P_{X=x} = P_{\mathcal{B}}$, $\omega \in X^{-1}(x)$.

Exercise 3.5.

(i) Let $\Omega = [0, 1]^2$ and let P be the Lebesgue measure on Ω. Let Δ be the class of all intersections $\Omega \cap \{(x, y) | x + y = a\}$, $0 \le a \le 2$. Determine the conditional probability measure $P_{\Delta, \xi}$, $\xi \in \Delta$.

(ii) $\Omega = \mathbb{R}^2$, $P =$ the two-dimensional distribution with density $f(x, y)$, and $X(\omega) = x$, $Y(\omega) = y$ for $\omega = (x, y)$. Prove that the conditional probability distribution of Y under $X = x$, $P^Y_{X=x}(E)$ $= P_{X=x} (Y \in E)$, is a one-dimensional distribution with density

$$\frac{f(x, y)}{\int_{\mathbb{R}^1} f(x, y)\, dy}.$$

(iii) Prove that if $\mathcal{B}_1$ and $\mathcal{B}_2$ are independent, then

$$P_{\mathcal{B}_1}(B_2) = P(B_2) \quad \text{a.s.,} \qquad B_2 \in \mathcal{B}_2.$$

[*Hint*: If $B_i \in \mathcal{B}_i$, $i = 1, 2$, then

$$E\big(P_{\mathcal{B}_1}(B_2), B_1\big) = P(B_2 \cap B_1) = P(B_2)P(B_1)$$
$$= E\big(P(B_2), B_1\big).]$$

(iv) Prove that $P_{\mathcal{D}(P)}(B) = 1_B(\omega)$ a.s. [*Hint*: $E(P_{\mathcal{D}(P)}(B), B_1) = P(B \cap B_1) = E(1_B, B_1)$, $B_1 \in \mathcal{D}(P)$.]

3.6 Properties of conditional probability measures

In the last section we introduced three kinds of conditional probability measures:

$$P_\Delta = P_{\Delta, \xi}, \qquad P_{X=x}, \qquad P_{\mathcal{B}} = P_{\mathcal{B}, \omega}.$$

They are closely related to each other. $P_{\Delta, \xi}$ is derived from $P_{X=x}$ by the relation:

$$P_{\Delta, \xi} = P_{X_\Delta = \xi} \qquad (X = X_\Delta).$$

Hence the properties of $P_{\Delta, \xi}$ can be derived from those of $P_{X=\xi}$. Since $P_{\mathcal{B}, \omega}$ was defined in terms of $P_{\Delta, \xi}$, we can derive the properties of $P_{\mathcal{B}, \omega}$ from those of $P_{\Delta, \xi}$ and hence from those of $P_{X=x}$. Therefore we will discuss only the properties of $P_{X=x}$.

Theorem 3.6.1. If X and Y are random variables taking values in S and T, respectively, then we have the following:

(i) $\quad P\{ X(\omega) \in A, \, Y(\omega) \in B \} = \int_A P^X(ds) P_{X=s}\{ Y \in B \}$

$$(A \in \mathscr{D}(P^X), \, B \in \mathscr{D}(P^Y)),$$

(ii) $\quad P\{(X(\omega), Y(\omega)) \in C \} = \int_S P^X(ds) P_{X=s}\{ Y \in C(s) \}$

$$(C \in \mathscr{D}(P^{(X,Y)}), \quad C(s) = \{ t \in T \mid (s,t) \in C \}).$$

Proof: Since $Y^{-1}(B) \in \mathscr{D}(P)$, we can apply Theorem 3.5 to $\Delta = \Delta_X$ to obtain

$$Y^{-1}(B) \cap \{ X=s \} \in \mathscr{D}(P_{X=s}), \qquad \text{a.e. } (P^X) \text{ on } S,$$

$$P\{ X^{-1}(A) \cap Y^{-1}(B) \} = \int_A P^X(ds) P_{X=s}\{ Y^{-1}(B) \} \qquad (A \in \mathscr{D}(P^X)).$$

This proves (i).

By the definition of random variables we have countable separating classes

$$\{ A_n \} \subset \mathscr{D}(P^X), \qquad \{ B_n \} \subset \mathscr{D}(P^Y)$$

on S and T, respectively. If C is a set of the form:

$$A_m \times B_n, \qquad m, n = 1, 2, \ldots,$$

then (ii) follows at once from (i). Let $\mathscr{C}$ denote the class of all such sets. Then the Dynkin class theorem ensures that (ii) holds for every $C \in \sigma[\mathscr{C}]$. $P^{(X,Y)}$ is the Lebesgue extension of its restriction to $\sigma[\mathscr{C}]$ by Theorem 3.1.2. Hence, for every $C \in \mathscr{D}(P^{(X,Y)})$ we can find $C_1, C_2 \in \sigma[\mathscr{C}]$ such that

$$C_1 \subset C \subset C_2, \qquad P^{(X,Y)}(C_2 - C_1) = 0.$$

Observing that (ii) holds for C_1 and C_2 and noting that

$$C_1(s) \subset C(s) \subset C_2(s),$$

we can see that (ii) holds for C. ∎

Since $P_{X=s}$ is a separable perfect probability measure, $(\Omega, P_{X=s})$ is a separable perfect probability space. Take a separating sequence $\{ B_n \} \subset \mathscr{D}(P^Y)$ on T,

$$Y^{-1}(B_n) \in \mathscr{D}(P_{X=s}) \qquad (n = 1, 2, \ldots; \, s \in S - N),$$

where N is a P^X-null set independent of n. Hence Y is regarded as a

T-valued random variable on $(\Omega, P_{X=s})$ for every $s \in S - N$ and as such we can define the probability distribution of Y on $(\Omega, P_{X=s})$, written $P^Y_{X=s}$; that is,

$$P^Y_{X=s}(E) = P_{X=s}(Y^{-1}(E)).$$

This is called the *conditional distribution* of Y under $X = s$. Using this notation, we can express formula (ii) as follows:

$$P^{(X,Y)}(C) = \int_S P^X(ds) P^Y_{X=s}(C(s)).$$

In case Y is a topological random variable on (Ω, P), it is so on $(\Omega, P_{X=s})$ for almost every (P^X) $s \in S$. If Y is a real random variable on (Ω, P), it is so on $(\Omega, P_{X=s})$ and as such its expectation (mean value) $E_{X=s}(Y)$ is defined for almost every (P^X) $s \in S$ whenever $E(Y)$ exists. This expectation is called the *conditional expectation* (*or mean*) of Y under $X = s$.

The following theorem, similar to Fubini's theorem for direct product measures, follows easily from the last theorem and the definition of integrals.

Theorem 3.6.2. If X and Y are random variables taking values in S and T, respectively, then

$$E\{f(X(\omega), Y(\omega))\} = \int_S P^X(ds) E_{X=s}(f(s, Y)).$$

Precisely speaking, if the left-hand side is well-defined, then so is the right-hand side and the equality holds.

Remark: For a decomposition Δ and a σ-algebra $\mathscr{B}$ we can define $P_{\Delta, \xi}$, $E_{\Delta, \xi}$, $P_{\mathscr{B}, \omega}$, and $E_{\mathscr{B}, \omega}$ in the same way as above.

Exercise 3.6.
   (i) Prove that if $E(f(X, Y))$ is well-defined, then

   $$E_{X=s}(f(X, Y)) = E_{X=s}(f(s, Y)) \quad \text{a.e. } (P^X) \text{ on } S.$$

   [*Hint*: $P_{X=s}$ is concentrated on $X^{-1}(s)$. a.e. (P^X).]
   (ii) Prove that if X and Y are independent, then

   $$P^Y_{X=s} = P^Y \quad \text{a.e. } (P^X)$$

   $$E_{X=s}(f(X, Y)) = E(s, Y) \quad \text{a.e. } (P^X)$$

   whenever $E(f(X, Y))$ is well-defined.

(iii) Prove that if X, Y, and Z are random variables taking values in S, T, and U, respectively, and if $E(f(X, Y, Z))$ is well-defined, then

$$E(f(X, Y, Z)) = \int_S P^X(ds) \int_T P^Y_{X=s}(dt) E_{(X, Y)=(s, t)}(f(s, t, Z)).$$

(iv) Let $(P_{X=s})_{Y=t}$ be the conditional probability distribution on Ω under $Y = t$ on the conditional probability space $(\Omega, P_{X=s})$. Prove that

$$(P_{X=s})_{Y=t} = P_{(X, Y)=(s, t)}$$

for almost every $(P^{(X, Y)})$ (s, t).

3.7   Real random variables

Let $L^0 = L^0(\Omega, P)$ denote the set of all real random variables. Two elements X and Y in L^0 are identified if $X(\omega) = Y(\omega)$ a.s. L^0 is a vector space with real coefficients when linear combinations are defined in the natural way.

We introduce a metric ρ_0 as follows:

$$\rho_0(X, Y) = E(|X - Y| \wedge 1) \qquad (a \wedge b = \min(a, b)).$$

Obviously it has the properties of a metric:

$$\rho_0(X, Y) = 0, \qquad \rho_0(X, Y) = \rho_0(Y, X),$$
$$\rho_0(X, Y) + \rho_0(Y, Z) \geq \rho_0(X, Z).$$

Also, it satisfies the separation axiom in the following sense:

$$\rho_0(X, Y) = 0 \Rightarrow X(\omega) = Y(\omega) \quad \text{a.s.}$$

Suppose that $X_1, X_2, \ldots, X \in L^0$. If $P\{|X_n - X| > \varepsilon\} \to 0$ $(n \to \infty)$ for every $\varepsilon > 0$, then X_n is said to *converge in probability* to X, written

$$X_n \to X \quad \text{i.p.} \quad (\text{i.p.} = \text{in probability})$$

$X_n \to X$ i.p. if and only if $\rho_0(X_n, X) \to 0$. This follows from the inequality:

$$\varepsilon P\{|X - Y| > \varepsilon\} \leq \rho_0(X, Y) \leq \varepsilon + P\{|X - Y| > \varepsilon\} \qquad (0 < \varepsilon < 1),$$

which is proved as follows:

$$\rho_0(X, Y) = E(|X - Y| \wedge 1) \geq E(\varepsilon, |X - Y| > \varepsilon)$$
$$\geq \varepsilon P\{|X - Y| > \varepsilon\},$$
$$\rho_0(X, Y) \leq E(\varepsilon, |X - Y| \leq \varepsilon) + E(1, |X - Y| > \varepsilon)$$
$$\leq \varepsilon + P\{|X - Y| > \varepsilon\}.$$

If $P\{X_n \to X\} = 1$, X_n is said to *converge almost surely* to X, written $X_n \to X$ a.s. Almost sure convergence implies convergence in probability, but the converse is not true, as the following example shows:

$$\Omega = [0,1), \qquad P = \text{the Lebesgue measure},$$

$$X_{n,k} = \text{the indicator of } \left[\frac{k-1}{n}, \frac{k}{n}\right), \qquad X \equiv 0,$$

$$\{X_{1,1}, X_{2,1}, X_{2,2}, X_{3,1}, X_{3,2}, X_{3,3}, \ldots\} = \{X_1, X_2, \ldots\},$$

$$X_n \to X \quad \text{i.p. but not a.s.}$$

If $X_n \to X$ i.p., then a subsequence of $\{X_n\}$ converges a.s. to X. By the assumption we can find a subsequence $\{Y_n\}$ such that

$$P\{|Y_n - X| > 2^{-n}\} < 2^{-n}.$$

Using the Borel–Cantelli lemma we can verify that

$$P\{|Y_n - X| \le 2^{-n} \text{ f.e.}\} = 1,$$

which implies that $Y_n \to X$ a.s.

Theorem 3.7.1. (L^0, ρ_0) is a complete separable metric space.

Proof: Let $\{X_n\}$ be a Cauchy sequence with respect to ρ_0; that is,

$$\rho_0(X_m, X_n) \to 0 \qquad (m, n \to \infty).$$

Then (X_n) has a subsequence $\{Y_n\}$ such that

$$\rho_0(Y_{n+1}, Y_n) < 2^{-n}, \qquad n = 1, 2, \ldots.$$

This implies that

$$E\left\{\sum_n |Y_{n+1} - Y_n| \wedge 1\right\} = \sum_n \rho_0(Y_{n+1}, Y_n) < \infty.$$

Hence

(1) $$\sum_n (|Y_{n+1} - Y_n| \wedge 1) < \infty \quad \text{a.s.}$$

and so $|Y_{n+1} - Y_n| \wedge 1 \to 0$ a.s., which implies that $|Y_{n+1} - Y_n| \wedge 1 = |Y_{n+1} - Y_n|$ (f.e.) a.s. This, combined with (1), implies that $\sum_n |Y_{n+1} - Y_n| < \infty$ a.s. Therefore Y_n converges a.s. to a random variable $Y \in L^0$, and so

(2) $$\rho_0(Y_n, Y) \to 0 \qquad (n \to \infty).$$

Since $\{Y_n\}$ is a subsequence of $\{X_n\}$, $\rho_0(X_m, X_n) \to 0$ $(m, n \to \infty)$ implies that

$$\rho_0(X_n, Y_n) \to 0 \qquad (n \to \infty).$$

This, together with (2), implies that

$$\rho_0(X_n, Y) \to 0 \qquad (n \to \infty),$$

which proves the completeness of (L^0, ρ_0).

P is strictly isomorphic to a K-regular probability measure μ on a set $S \subset \mathbb{R}^1$. To prove the separability of (L^0, ρ_0) we can assume without any loss of generality that $(\Omega, P) = (S, \mu)$. Then the functions

$$\sum_{i=1}^{n} a_i 1_{[\alpha_i, \beta_i]}, \qquad a_i, \alpha_i, \beta_i \in Q$$

form a countable dense subset of (L^0, ρ_0). Hence (L^0, ρ_0) is separable. ∎

It is well-known in functional analysis that

$$L^p = L^p(\Omega, P) = \{ X \in L^0 | E(|X|^p) < \infty \} \qquad (1 \le p < \infty)$$

is a Banach space with norm $\|X\|_p = (E(|X|^p))^{1/p}$.

Since the probability distribution of X, P^X, is a one-dimensional distribution, $(|M|_p(P^X))^{1/p}$ increases with p by Theorem 2.5.4. Since

$$|M|_p(P^X) = \int_{R^1} |x|^p P^X(dx) = \int_\Omega |X(\omega)|^p P(d\omega) = E(|X|^p)$$

holds by the formula of change of variables, $\|X\|_p = (E(|X|^p))^{1/p}$ increases with p. (This fact can also be proved directly in the same way as Theorem 2.5.4.) Hence it holds that if $p < q$, then

$$L^q \subset L^p,$$
$$\|X_n - X\|_q \to 0 \Rightarrow \|X_n - X\|_p \to 0.$$

Observing that

$$E(|X|^p) \ge E(|X|^p, |X| \ge \varepsilon) \ge \varepsilon^p P(|X| \ge \varepsilon),$$

we see that

$$\|X_n - X\|_p \to 0 \Rightarrow X_n \to X \text{ i.p.}$$

If $\|X_n - X\|_p \to 0$, X_n is said to *converge in the pth order mean* to X. Since convergence in mean implies convergence i.p., every sequence convergent in mean has an a.s. convergent subsequence.

If the probability distribution of X_n converges to that of X in the sense mentioned in Section 2.5.b, X_n is said to *converge in distributions* to X. This convergence is meaningful, even though $X_1, X_2, \ldots, X$ are defined on different probability spaces. Since

$$E(f(Y)) = \int_{\mathbb{R}^1} f(y) P^Y(dy)$$

hold by the formula of change of variables, Theorem 2.5.2 ensures that X_n converges in distributions if and only if

$$E(g(X_n)) \to E(g(X))$$

whenever $g(x)$ is continuous and has compact support.

Theorem 3.7.2. If $X_n \to X$ i.p., then X_n converges in distribution to X.

Proof: Let g be a continuous real function with compact support. Then g is uniformly continuous and bounded and

$$|E(g(X_n)) - E(g(X))| \leq E(|g(X_n) - g(X)|)$$
$$\leq E(|g(X_n) - g(X)|, |X_n - X| \leq \varepsilon)$$
$$+ 2AP\{|X_n - X| > \varepsilon\} \qquad \left(A = \sup_x |g(x)|\right)$$
$$\leq \sup\{|g(x) - g(y)| \mid |x - y| < \varepsilon\}$$
$$+ 2AP\{|X_n - X| > \varepsilon\}.$$

Observing that $P\{|X_n - X| > \varepsilon\} \to 0 \ (n \to \infty)$ by the assumption, we obtain

$$\limsup_{n \to \infty} |E(g(X_n)) - E(g(X))| \leq \sup\{|g(x) - g(y)| \mid |x - y| < \varepsilon\}$$
$$\to 0 \qquad (\varepsilon \to 0).$$

This implies that $E(g(X_n)) \to E(g(X)) \ (n \to \infty)$. ∎

Summarizing the relation among the different types of convergence mentioned above, we have

> a.s. convergence $\Rightarrow$
> mean convergence $\Rightarrow$ convergence i.p.
> $\Rightarrow$ convergence in distribution.

If $\|X_n - X\|_1 \to 0$, we can find a subsequence of X_n converging to X a.s. But the converse is not true. Even though $X_n \in L^1$ converges to X a.s., it does not always hold that $\|X_n - X\|_1 \to 0$. Now the following problem arises: Under what condition can we conclude $\|X_n - X\|_1 \to 0$ from $X_n \to X$ a.s.?

To answer this we introduce the concept of uniform integrability. A family of real random variables $\{X_\lambda, \lambda \in \Lambda\} \subset L^1$ is called *uniformly integrable* if

$$\sup_\lambda E(|X_\lambda|, |X_\lambda| > A) \to 0 \qquad (A \to \infty).$$

Since X_λ is integrable, it is obvious that for each λ we have

$$E(|X_\lambda|, |X_\lambda| > A) \to 0 \qquad (A \to \infty).$$

The uniform integrability asserts that this convergence is uniform in $\lambda \in \Lambda$.

Theorem 3.7.3. If $\{ X_\lambda, \lambda \in \Lambda \}$ is uniformly integrable, then
    (i) $\|X_\lambda\|_1$ $(\lambda \in \Lambda)$ is bounded,
    (ii) $\sup_\lambda P\{|X_\lambda| > A\} \to 0$ $(A \to \infty)$,
    (iii) $P(B) \to 0 \Rightarrow \sup_\lambda E(|X_\lambda| B) \to 0$.
   Either (ii) and (iii) or (i) and (iii) is necessary and sufficient for the uniform integrability of $\{ X_\lambda, \lambda \in \Lambda \}$.

Proof:
    (i) It suffices to observe that

$$\|X_\lambda\| = E(|X_\lambda|, |x_\lambda| > A) + E(|X_\lambda|, |X_\lambda| \leq A)$$

$$\leq E(|X_\lambda|, |X_\lambda| > A) + A.$$

    (ii) This follows from (i), because

$$\|X_\lambda\|_1 \geq E(|X_\lambda|, |X_\lambda| > A) \geq AP\{|X_\lambda| > A\}$$

    (iii) Taking $A = A(\varepsilon)$ sufficiently large for $\varepsilon > 0$, we have

$$\sup_\lambda E(|X_\lambda|, |X_\lambda| > A) < \varepsilon.$$

If $P(B) < \varepsilon/A$, then

$$E(|X_\lambda|, B) \leq E(|X_\lambda|, |X_\lambda| > A) + E(|X_\lambda|, \{|X_\lambda| \leq A\} \cap B)$$

$$< \varepsilon + AP(B) < 2\varepsilon.$$

Thus the first part of the theorem is proved.

To prove that (ii) and (iii) imply the uniform integrability, it suffices to put $B = \{|X_\lambda| > A\}$ in (iii) in view of (ii). Since (i) implies (ii), (i) and (iii) also imply the uniform integrability. Hence the second part of the theorem is proved. ∎

Theorem 3.7.4. Suppose that $X_1, X_2, \ldots, X \in L^1$ and $X_n \to X$ a.s. Then $\|X_n - X\|_1 \to 0$ if and only if $\{ X_n \}$ is uniformly integrable.

Proof: Suppose that $\|X_n - X\|_1 \to 0$. Then $\|X_n\|_1$ is bounded, because $\|X_n\|_1 \le \|X\|_1 + \|X_n - X\|_1$. Since X_n and X are integrable, we find, for every $\varepsilon > 0$, $\delta, \delta_1, \delta_2, \ldots$ such that

$$P(B) < \delta \Rightarrow E(|X|, B) < \varepsilon,$$
$$P(B) < \delta_n \Rightarrow E(|X_n|, B) < \varepsilon.$$

Observing that

$$E(|X_n|, B) \le E(|X|, B) + E(|X_n - X|, B)$$
$$\le E(|X|, B) + \|X_n - X\|_1,$$

and taking, for every $\varepsilon > 0$, n_0 sufficiently large so that $\|X_n - X\|_1 < \varepsilon$ for every $n > n_0$, we obtain

$$P(B) < \delta \Rightarrow E(|X_n|, B) < 2\varepsilon \qquad (n = n_0 + 1, n_0 + 2, \ldots)$$

and so

$$P(B) < \min(\delta, \delta_1, \ldots, \delta_{n_0}) \Rightarrow \forall n \; E(|X_n|, B) < 2\varepsilon.$$

Thus the assertions (i) and (iii) of Theorem 3.7.3 are verified for $\{X_n\}$. Hence $\{X_n\}$ is uniformly integrable by the second part of the theorem.

Suppose conversely that $\{X_n\}$ is uniformly integrable. Using (ii) of the last theorem we can check that as $A \to \infty$,

$$P(\{|X_n| > A\} \cup \{|X| > A\}) \le P(|X_n| > A) + P(|X| > A) \to 0$$

uniformly in n. Denote by $B_{n,A}$ the ω-set in $\{\ \}$ of the left-hand side. Then

$$\sup_n P(B_{n,A}) \to 0 \qquad (A \to \infty).$$

Also, we have

$$\|X_n - X\|_1 = E(|X_n - X|, B_{n,A}) + E(|X_n - X|, \{|X_n| \le A\} \cap \{|X| \le A\})$$
$$\le E(|X_n|, B_{n,A}) + E(|X|, B_{n,A}) + E(|X_n - X| \wedge 2A).$$

Hence we can use (iii) of the last theorem to find, for every $\varepsilon > 0$, $A_1(\varepsilon)$ such that the first term in the last expression is less than ε for every n whenever $A \ge A_1(\varepsilon)$. Since X is integrable, we can find $A_2(\varepsilon)$ such that the second term is less than ε for every n whenever $A \ge A_2(\varepsilon)$. For $A = \max(A_1(\varepsilon), A_2(\varepsilon))$ the last term converges to 0 as $n \to \infty$ by the bounded convergence theorem, because $X_n \to X$ a.s. Hence

$$\limsup_{n \to \infty} \|X_n - X\|_1 \le 2\varepsilon \to 0 \; (\varepsilon \downarrow 0). \qquad \blacksquare$$

Let X be a real random variable. Then its probability distribution P^X is a one-dimensional distribution. The distribution function of P^X defined in Section 2.5 is called the *distribution function* of X, written $F_X(x)$. Similarly, we define *the pth order moment* $M_p(X)$, the *pth order absolute moment* $|M|_p(X)$, the *characteristic function* $\varphi_X(X)$, the *variance* $V(X)$, and the *standard deviation* $\sigma(X)$. The formula of change of variables implies that

$$\int_{\mathbb{R}^1} f(x) P^x(dx) = E(f(X)).$$

Setting $f(x) = 1_{(-\infty, x)}, x^p, |x|^p, e^{izx}$, we obtain

$$F_X(x) = P\{X \leq x\}, \qquad M_p(X) = E(X^p),$$

$$|M|_p(X) = E(|X|^p), \qquad \varphi_X(z) = E(e^{izX}),$$

$$V(X) = E((X - EX)^2), \qquad \sigma(X) = \sqrt{V(X)}.$$

If $X \in L^2$, then $\sigma(X) < \infty$ and we have *Chebyshev's inequality*:

$$P\{|X - EX| \geq a\sigma(X)\} \leq 1/a^2 \qquad (a > 0; \sigma(X) > 0),$$

because

$$\sigma(X)^2 \geq E((X - EX)^2, |X - EX| \geq a\sigma(X)),$$

$$\geq a^2\sigma(X)^2 P\{|X - EX| \geq a\sigma(X)\}.$$

If $X, Y \in L^2$, the covariance of X and Y, written $V(X, Y)$, is defined as follows:

$$V(X, Y) = E((X - EX)(Y - EY))$$

$$= \iint_{\mathbb{R}^2} (x - EX)(y - EY) P^{(X,Y)}(dx\, dy)$$

If $0 < \sigma(X), \sigma(Y) < \infty$, then

$$R(X, Y) = \frac{V(X, Y)}{\sigma(X)\sigma(Y)}$$

is called the *correlation coefficient* of X and Y.

If $V(X_i) < \infty$ $(i = 1, 2, \ldots, n)$ then the additivity of mean values implies

(A) $\qquad V\left(a_0 + \sum_{i=1}^n a_i X_i\right) = \sum_{i,j=1}^n a_i a_j V(X_i, X_j) \qquad (a_0, a_1, \ldots, a_n: \text{real}).$

Theorem 3.7.5. Suppose that $X_1, X_2, \ldots, X_n$ are independent. Then we have the following identities:

(i) $V(X_i, X_j) = 0 \qquad (i \neq j)$, if $X_i \in L^2$,

(ii) $V(a_0 + \sum_{i=1}^n a_i X_i) = \sum_{i=1}^n a_i^2 V(X_i)$, if $X_i \in L^2$,

(iii) $\varphi_{X_1 + X_2 + \cdots + X_n}(z) = \varphi_{X_1}(z)\varphi_{X_2}(z)\ldots\varphi_{X_n}(z)$,

(iv) $P_{X_1 + X_2 + \cdots + X_n} = P_{X_1} * P_{X_2} * \cdots * P_{X_n}.$

Proof: Identities (i) and (iii) are obvious by the multiplicativity of mean values for independent random variables (Theorem 3.4.6). Identity (ii) follows from (i) and (A). Using (iii), we obtain

$$\mathscr{F}P_{X_1 + X_2 + \cdots + X_n}(z) = \varphi_{X_1 + X_2 + \cdots + X_n}(z) = \varphi_{X_1}(z)\varphi_{X_2}(z)\ldots\varphi_{X_n}(z)$$

$$= \prod_{i=1}^{n} \mathscr{F}P_{X_i}(z) = \mathscr{F}(P_{X_1} * P_{X_2} * \cdots * P_{X_n})(z),$$

which implies (iv), because the correspondence $\mu \leftrightarrow \mathscr{F}\mu$ is 1-to-1. ∎

For a random vector $X = (X_1, X_2, \ldots, X_n)$ we define the following:

$$EX = (EX_1, EX_2, \ldots, EX_n), \qquad \text{mean vector}$$

$$V(X) = (V(X_i, X_j))_{i,j=1}^{n}, \qquad \text{variance matrix}$$

$$\varphi(z) = E(e^{i(z, X)}) \qquad \text{characteristic function}$$

$$\left(z = (z_1, z_2, \ldots, z_n), \qquad (z, X) = \sum_{i=1}^{n} z_i X_i\right)$$

Theorem 3.7.6 (Kac's theorem). $X_1, X_2, \ldots, X_n$ are independent if and only if

$$E\left(\exp\left(i \sum_{k=1}^{n} z_k X_k\right)\right) = \prod_{i=1}^{n} E(e^{iz_k X_k})$$

holds for every $z = (z_1, z_2, \ldots, z_n) \in \mathbb{R}^n$.

Proof: If $X_1, X_2, \ldots, X_n$ are independent, then the multiplicativity of mean values implies the equality. Suppose conversely that the equality holds. The left-hand side is equal to

$$(\mathscr{F}P^X)(z) \qquad (X = (X_1, X_2, \ldots, X_n))$$

by the formula of change of variables, and the right-hand side is equal to

$$\prod_{i=1}^{n} \int_{\mathbb{R}^1} e^{iz_k x_k} \mu_k(dx_k) \qquad (\mu_k = P^{X_k})$$

$$= \int\int \cdots \int_{\mathbb{R}^n} e^{i(z, x)} (\mu_1 \times \mu_2 \times \cdots \times \mu_n)(dx_1 \, dx_2 \ldots dx_n)$$

$$= \mathscr{F}(\mu_1 \times \mu_2 \times \cdots \times \mu_n)(z)$$

by Fubini's theorem. Hence the assumption asserts that

$$(\mathscr{F}P^X)(Z) = \mathscr{F}(\mu_1 \times \mu_2 \times \cdots \times \mu_n).$$

Since $\mu \leftrightarrow \mathscr{F}\mu$ is 1-to-1 in $\mathbb{R}^n$, this implies

$$P^X = \mu_1 \times \mu_2 \times \cdots \times \mu_n.$$

Therefore, we have

$$P(X_1 \in E_1, X_2 \in E_2, \ldots, X_n \in E_n)$$
$$= P^X(E_1 \times E_2 \times \cdots \times E_n)$$
$$= \mu_1(E_1)\mu_2(E_2)\ldots\mu_n(E_n) = \prod_{k=1}^{n} P(X_k \in E_k),$$

which implies the independence of $X_1, X_2, \ldots, X_n$. ■

Exercise 3.7.

(i) Let $p > 1$. Prove that if $|M|_p(X_\lambda)$ $(\lambda \in \Lambda)$ is bounded, then $\{X_\lambda, \lambda \in \Lambda\}$ is uniformly integrable. [*Hint*: If $|X_\lambda| \geq A$, then $|X_\lambda|^p \geq |X_\lambda|A^{p-1}$. Hence

$$E(|X_\lambda|, |X_\lambda| \geq A) \leq E(|X_\lambda|^p A^{1-p}, |X_\lambda| \geq A) \leq A^{1-p}E(|X_\lambda|^p).]$$

(ii) Give an example of $\{X_n\}$ satisfying the following conditions:
(a) $X_n \to X$ a.s., $X_n \in L^1$ $(n = 1, 2, \ldots)$ but not $X \in L^1$.
(b) $X_n \to X$ a.s., $X_n \in L^1$ $(n = 1, 2, \ldots)$, $X \in L^1$ but not $\|X_n - X\|_1 \to 0$.
[*Hint*: $\Omega = (0,1)$, P = the Lebesgue measure. Observe the following:

$$X_n(\omega) \equiv \omega^{-(n-1)/n} \quad \text{and} \quad X_n(\omega) = n1_{(0,1/n)}(\omega).]$$

(iii) Prove that if both $\{X_\lambda, \lambda \in \Lambda\}$ and $\{Y_\lambda, \lambda \in \Lambda\}$ are uniformly integrable, then $\{X_\lambda + Y_\lambda, \lambda \in \Lambda\}$ is uniformly integrable. [*Hint*: Use the second part of Theorem 3.7.3.]

(iv) (*Bienaymé's inequality.*) Prove that if $f: \{0, \infty) \to [0, \infty)$ is an increasing (= nondecreasing) function, then

$$P(|X| \geq a) \leq \frac{1}{f(a)}E(f(|X|)) \quad (f(a) > 0).$$

[*Hint*:

$$E(f(|X|)) \geq E(|f(X)|, |X| \geq a) \geq f(a)P(|X| \geq a).]$$

(v) Let μ_n be the probability distribution of X_n. Prove that

$$X_n \to 0 \text{ i.p.} \Leftrightarrow \mu_n \to \delta.$$

[*Hint*: $\Rightarrow$ is obvious. To prove $\Leftarrow$, use Theorem 2.5.2((i) $\Leftrightarrow$ (iv)) and $P\{|X_n| \geq \varepsilon\} = \mu_n((-\varepsilon, \varepsilon)^c).]$

3.8   Conditional mean operators

Definition 3.8. The map from $L^1 = L^1(\Omega, P)$ into itself carrying $Y \in L^1$ to the $Y_{\mathscr{B}} \in L^1$ defined by

(1) $$Y_{\mathscr{B}}(\omega) = E_{\mathscr{B},\omega}(Y) = \int_{\Omega} Y(\omega') P_{\mathscr{B},\omega}(d\omega')$$

is called the *conditional mean operator* relative to $\mathscr{B}$, written $E_{\mathscr{B}}$.

To verify that this definition is meaningful we have to prove that $Y_{\mathscr{B}} \in L^1$ for $Y \in L^1$. For this purpose we will briefly review the definition of $P_{\mathscr{B},\omega}$ (Section 3.5). Since P is a separable perfect measure, we can find a separable decomposition such that $\mathscr{B}_\Delta$ is equivalent to $\mathscr{B}$; that is,

$$\mathscr{B}_\Delta \vee \mathbf{2} = \mathscr{B} \vee \mathbf{2}.$$

$P_{\mathscr{B},\omega}$ is defined by

(2) $$P_{\mathscr{B},\omega} = P_{\Delta, X_\Delta(\omega)}, \quad X_\Delta(\omega) = \text{the element of } \Delta \text{ containing } \omega.$$

It is easy to see that $P_{\mathscr{B},\omega}$ is independent of the choice of Δ. If $\mathscr{B}_1$ is equivalent to $\mathscr{B}_2$, then $P_{\mathscr{B}_1} = P_{\mathscr{B}_2}$ a.s.

The conditional probability measure under a random variable, $P_{X=s}$, is related to that under a σ-algebra as follows:

(3) $$P_{X=s}|_{s=X(\omega)} = P_{\bar{\sigma}[X],\omega},$$

where the left-hand side indicates the value of the function (of s) $P_{X=s}$ evaluated at $s = X(\omega)$ and is often denoted by $P(\cdot | X)$ or P_X. In case X is a topological random variable, $\bar{\sigma}[X]$ may be replaced by $\sigma[X]$.

It should be noted that $P_{\mathscr{B},\omega}$, as a function of ω, has different versions that coincide with each other a.s. As usual we fix a version of $P_{\mathscr{B},\omega}$ unless stated otherwise. As the probability measure $P_{\mathscr{B},\omega}$ is separable and perfect for each ω, so is the probability space $(\Omega, P_{\mathscr{B},\omega})$. Hence all concepts introduced on (Ω, P) can be considered on $(\Omega, P_{\mathscr{B},\omega})$ for each ω. For example, the conditional mean value of an extended real random variable on $(\Omega, P_{\mathscr{B},\omega})$, written $E_{\mathscr{B},\omega}(Y)$, is defined as follows:

$$E_{\mathscr{B},\omega}(Y) = \int_{\Omega} Y(\omega') P_{\mathscr{B},\omega}(d\omega')$$

whenever the integral is meaningful. Since the definition of $E_{\mathscr{B},\omega}(Y)$ is the same as that of $E(Y)$ except with P being replaced by $P_{\mathscr{B},\omega}$, all the properties of mean values hold for conditional mean values.

In accordance with different versions of $P_{\mathscr{B},\omega}$ we have different versions of $E_{\mathscr{B},\omega}$ that coincide with each other a.s. Even if we fix a version, the domain of definition of $E_{\mathscr{B},\omega}$ (the set of all Y's for which $E_{\mathscr{B},\omega}$ is defined) varies as ω moves on Ω.

Lemma 3.8.1. If $Y(\omega)$ is an extended real random variable taking values in $[0, \infty]$, then

    (i) $E_{\mathscr{B},\omega}(Y)$ is measurable $\mathscr{B} \vee \mathbf{2}$ (and so P-measurable) in ω,

    (ii) $E(E_{\mathscr{B},\omega}(Y), B) = E(Y, B)$, $B \in \mathscr{B}$.

Proof: First we consider the case where $Y = 1_A$ ($A \in \mathscr{D}(P)$). In this case we have

$$E_{\mathscr{B},\omega}(Y) = P_{\mathscr{B},\omega}(A) = P_{\mathscr{B}_\Delta,\omega}(A) = P_{\Delta, X_\Delta(\omega)}(A) \quad \text{a.s.}$$

by part (2) of definition 3.8. Since $P_{\Delta,\xi}(A)$ is P^Δ-measurable in ξ (Section 3.5) and $X_\Delta(\omega)$ is measurable $\mathscr{B}_\Delta$, $E_{\mathscr{B},\omega}(Y)$ is $\mathscr{B}_\Delta$-measurable. Since $\mathscr{B}_\Delta \subset \mathscr{B} \vee \mathbf{2}$, $E_{\mathscr{B},\omega}(Y)$ is measurable $\mathscr{B} \vee \mathbf{2}$ in ω and so P-measurable in ω. This proves assertion (i). To prove (ii) we can assume that $B \in \mathscr{B}_\Delta$, because $\mathscr{B}_\Delta$ is equivalent to $\mathscr{B}$. Since every $B \in \mathscr{B}_\Delta$ is expressible as $B = X_\Delta^{-1}(E)$ ($E \in \mathscr{D}(P^\Delta)$), we obtain

$$E(P_{\mathscr{B},\omega},(A), B) = E(P_{\mathscr{B}_\Delta,\omega}(A), B) = E\left(P_{\Delta, X_\Delta(\omega)}(A), X_\Delta^{-1}(E)\right)$$

$$= \int_E P_{\Delta,\xi}(A) P^\Delta(d\xi) = P(A \cap X^{-1}(E)) = P(A \cap B),$$

which proves (ii).

To discuss the general case it suffices to take linear combinations and limits as we always do in integration theory. ∎

We define $L^0_{\mathscr{B}}$ to be the set of all real random variables that are measurable $\mathscr{B}$ and define $L^p_{\mathscr{B}}$ to be $L^p \cap L^0_{\mathscr{B}}$ for $p \in [1, \infty)$. For $Y \in L^0$ we define $\|Y\|_\infty$ by

$$\|Y\|_\infty = \underset{\omega \in \Omega}{\text{ess.sup}} \, |Y(\omega)| = \inf\{b \mid |Y(\omega)| \leq b \quad \text{a.s.}\}.$$

Then the space

$$L^\infty = L^\infty(\Omega, P) = \{Y \in L^0 : \|Y\|_\infty < \infty\}$$

is a Banach space with norm $\| \ \|_\infty$, and

$$L_{\mathscr{B}}^\infty = L^\infty \cap L_{\mathscr{B}}^0$$

is a Banach subspace of L^∞. $L_{\mathscr{B}}^p$ $(1 \leq p < \infty)$ is also a closed subspace of L^p. L^p $(1 \leq p < \infty)$ is separable, because (Ω, P) is separable, but L^∞ is not in general, as the following example shows:

$$\Omega = [0, 1), \qquad P = \text{the Lebesgue measure}$$

$$\|1_{[\alpha, \beta)} - 1_{[\gamma, \delta)}\|_\infty = 1 \qquad \text{if } [\alpha, \beta) \neq [\gamma, \delta)$$

Two numbers $p, q \in [1, \infty]$ are called *conjugate* to each other if

$$p^{-1} + q^{-1} = 1 \qquad (\infty^{-1} = 0 \text{ by convention}).$$

Using the same method as in the proof of Lemma 2.5 we obtain

Hölder's inequality.

$$\|X \cdot Y\|_1 \leq \|X\|_p \|Y\|_q \qquad (p, q: \text{conjugate}).$$

Lemma 3.8.2. If $Y \in L^1$, then
   (i) $E_{\mathscr{B}, \omega}(Y)$ (as a function of ω) $\in L_{\mathscr{B}}^1$,
   (ii) $E(E_{\mathscr{B}, \omega}(Y), B) = E(Y, B)$ for every $B \in \mathscr{B}$.

Proof: Applying the last lemma to Y^+ and Y^- and observing that $E_{\mathscr{B}, \omega}(Y^\pm) < \infty$ follows from

$$E\left(E_{\mathscr{B}, \omega}(Y^\pm)\right) = E(Y^\pm) \leq E(|Y|) < \infty,$$

we can easily prove this lemma. ∎

Since two equivalent random variables are identified in L^p,

$$L_{\mathscr{B}_1}^p = L_{\mathscr{B}_2}^p \qquad \text{if } \mathscr{B}_1 \text{ is equivalent to } \mathscr{B}_2$$

and so

$$L_{\mathscr{B}}^p = L_{\mathscr{B} \vee 2}^p.$$

This lemma ensures that the operator $E_{\mathscr{B}}: L^1 \to L^1$ in Definition 3.8 is well-defined.

Theorem 3.8.1. Let $X \in L^1$. Then $\tilde{X} = E_{\mathscr{B}} X$ a.s. if and only if the following two conditions hold:
   (i) $\tilde{X} \in L_{\mathscr{B}}^1$,
   (ii) $E(\tilde{X}, B) - E(X, B), \qquad B \in \mathscr{B}.$

Proof: If $\tilde{X} = E_{\mathscr{B}} X$ a.s., then these conditions hold obviously. Suppose that these conditions are satisfied. Let

$$X' = E_{\mathscr{B}} X.$$

Then $X' \in L_{\mathscr{B}}^{\mathcal{L}}$ and $E(X', B) = E(X, B)$, $B \in \mathscr{B}$. Hence

$$E(\tilde{X}, B) = E(X', B), \qquad B \in \mathscr{B}.$$

Since both $\tilde{X}$ and X' are measurable $\mathscr{B}$, the set $B_0 = \{\tilde{X}(\omega) > X'(\omega)\} \in \mathscr{B}$. If $P(B_0) > 0$, then

$$E(\tilde{X}, B_0) > E(X', B_0),$$

which is a contradiction. Hence

$$P(B_0) = 0; \qquad \text{that is, } \tilde{X} \leq X' \quad \text{a.s.}$$

Similarly for the opposite inequality; and so we obtain

$$\tilde{X} = X' \quad \text{a.s.,}$$

which completes the proof. ∎

Theorem 3.8.2 (Jensen's inequality). Suppose that

$$X \in L^1(\Omega, P) \qquad \text{and} \qquad X(\Omega) \subset I,$$

where I is a real interval–open, semi-open, or closed and finite or infinite. If $f: I \rightarrow \mathbb{R}$ is convex, then

$$f(EX) \leq E(f(x)),$$
$$f(E_{\mathscr{B}} X) \leq E_{\mathscr{B}}(f(X)) \text{ a.s.}$$

The right-hand sides may be infinite.

Proof: Since $X(\Omega) \subset I$, $m \equiv EX \in I$. By convexity of f we can find α such that

$$f(x) \geq f(m) + \alpha(x - m), \qquad x \in I.$$

Hence

$$(1) \qquad f(X(\omega)) \geq f(m) + \alpha(X(\omega) - m), \qquad \omega \in \Omega.$$

As X belongs to L^1, so does the right-hand side. This implies that

$$E\{(f(X))^-\} \leq E(|f(m) + \alpha(X - m)|) < \infty,$$

which ensures that $E(f(X))$ is well-defined and takes a value in $(-\infty, \infty]$. Taking the mean values of both sides of (1) we obtain

$$E(f(X)) \geq f(m) = f(EX),$$

which proves the first inequality of the theorem.

Since

$$E\left(E_{\mathscr{B}}(|X|)\right) = E|X| < \infty,$$

we obtain

$$E_{\mathscr{B}}|X| < \infty \quad \text{a.s.};$$

that is,

$$X \in L^1(\Omega, P_{\mathscr{B}}) \quad \text{a.s.} \quad (P_{\mathscr{B}} = P_{\mathscr{B}, \omega}).$$

Replacing P by $P_{\mathscr{B}}$ in the first inequality we obtain the second inequality. ∎

Setting $I = (-\infty, \infty)$, $f(x) = |x|^p$ ($1 \leq p < \infty$) in this theorem, we can check that if $X \in L^p$ ($\subset L^1$), then

$$|E_{\mathscr{B}}X|^p \leq E_{\mathscr{B}}|X|^p,$$

$$E\left(|E_{\mathscr{B}}X|^p\right) \leq E\left(E_{\mathscr{B}}|X|^p\right) = E\left(|X|^p\right) < \infty.$$

Hence we obtain

$$X \in L^p \Rightarrow E_{\mathscr{B}}X \in L^p_{\mathscr{B}}, \qquad \|E_{\mathscr{B}}X\|_p \leq \|X\|_p$$

for $1 \leq p < \infty$. This holds also for $p = \infty$ obviously. Therefore $E_{\mathscr{B}}(L^p) \subset L^p_{\mathscr{B}}$. If $X \in L^p_{\mathscr{B}}$, then $\tilde{X} = X$ satisfies the conditions in Theorem 3.8.1, and so we obtain $E_{\mathscr{B}}X = X$. Hence $E_{\mathscr{B}}(L^p) = L^p_{\mathscr{B}}$.

Theorem 3.8.3. Suppose that p and q are conjugate, and $X \in L^p$. Then $\tilde{X} = E_{\mathscr{B}}X$ if and only if the following two conditions hold
  (i) $\tilde{X} \in L^p_{\mathscr{B}}$,
  (ii) $E(\tilde{X}Y) = E(XY)$, $\qquad Y \in L^q_{\mathscr{B}}$.

Proof: If these conditions hold, then the conditions of Theorem 3.8.1 follow by setting $Y = 1_B$ ($B \in \mathscr{B}$), so we obtain $\tilde{X} = E_{\mathscr{B}}X$. Suppose conversely that $\tilde{X} = E_{\mathscr{B}}X$. Then $\tilde{X}$ satisfies (i), as we proved above. Condition (ii) is valid for $Y = 1_B$ ($B \in \mathscr{B}$) by Theorem 3.8.1 and so for every linear combination of such Y's. But every $Y \in L^q_{\mathscr{B}}$ can be approximated by such linear combinations with respect to the norm $\| \|_q$. Hence we can use Hölder's inequality to check that condition (ii) holds for every $Y \in L^q_{\mathscr{B}}$. ∎

Theorem 3.8.4: Suppose that p and q are conjugate and $X \in L^p$, $Y \in L^q$. Then
  (i) $E((E_{\mathscr{B}}X)Y) = E(X(E_{\mathscr{B}}Y)) = E((E_{\mathscr{B}}X)(E_{\mathscr{B}}Y))$.
  (ii) $Y \in L^q_{\mathscr{B}} \Rightarrow E_{\mathscr{B}}(XY) = YE_{\mathscr{B}}X$.

Proof.

(i) Since $E_{\mathscr{B}}Y \in L_{\mathscr{B}}^q$, we can use the last theorem (ii) to check that $E(XE_{\mathscr{B}}Y) = E((E_{\mathscr{B}}X)(E_{\mathscr{B}}Y))$. Similarly, we obtain $E(YE_{\mathscr{B}}X) = E((E_{\mathscr{B}}Y)(E_{\mathscr{B}}X))$.

(ii) Since $X \in L^p$, $Y \in L_{\mathscr{B}}^q \subset L^q$ by the assumption, we can use Hölder's inequality to check that $XY \in L^1$, $YE_{\mathscr{B}}X \in L_{\mathscr{B}}^1$, and $YZ \in L_{\mathscr{B}}^q$ for every $Z \in L_{\mathscr{B}}^\infty$. Hence the last theorem (the "only if" part) ensures that $E(XYZ) = E((E_{\mathscr{B}}X)YZ)$ for every $Z \in L_{\mathscr{B}}^\infty$. By the last theorem (the "if" part, $p = 1$, $q = \infty$) we obtain $E_{\mathscr{B}}(XY) = (E_{\mathscr{B}}X)Y$. ∎

Theorem 3.8.5. If $X \in L^2$, then $\tilde{X} = E_{\mathscr{B}}X$ is characterized as the element $\tilde{X}$ in $L_{\mathscr{B}}^2$ minimizing $\|X - \tilde{X}\|_2$. Hence $E_{\mathscr{B}}$: $L^2 \to L^2$ is the orthogonal projection from the Hilbert space L^2 to its subspace $L_{\mathscr{B}}^2$.

Proof: For every $\tilde{X} \in L_{\mathscr{B}}^2$ we have

$$\|X - \tilde{X}\|_2^2 = \|(X - E_{\mathscr{B}}X) - (\tilde{X} - E_{\mathscr{B}}X)\|_2^2$$

$$= \|X - E_{\mathscr{B}}X\|_2^2 + 2E((X - E_{\mathscr{B}}X)(\tilde{X} - E_{\mathscr{B}}X)) + \|\tilde{X} - E_{\mathscr{B}}X\|_2^2,$$

$$X - E_{\mathscr{B}}X \in L^2, \qquad \tilde{X} - E_{\mathscr{B}}X \in L_{\mathscr{B}}^2, \qquad E_{\mathscr{B}}(X - E_{\mathscr{B}}X) = 0.$$

But

$$E((X - E_{\mathscr{B}}X)(\tilde{X} - E_{\mathscr{B}}X)) = E((E_{\mathscr{B}}(X - E_{\mathscr{B}}X))(\tilde{X} - E_{\mathscr{B}}X)) = 0$$

by Theorem 3.8.3(ii). Hence

$$\|X - \tilde{X}\|_2^2 = \|X - E_{\mathscr{B}}X\|_2^2 + \|\tilde{X} - E_{\mathscr{B}}X\|_2^2 \qquad (\tilde{X} \in L_{\mathscr{B}}^2)$$

$$\geq \|X - E_{\mathscr{B}}X\|_2^2$$

holds for every $\tilde{X} \in L_{\mathscr{B}}^2$ and the equality holds if and only if $\tilde{X} = E_{\mathscr{B}}X$ ($\in L_{\mathscr{B}}^2$). ∎

Theorem 3.8.6. If $\mathscr{B}_1 \subset \mathscr{B}_2$, then $E_{\mathscr{B}_1}E_{\mathscr{B}_2} = E_{\mathscr{B}_1}$ on L^p ($1 \leq p \leq \infty$).

Proof: Suppose that $X \in L^p$. Then $E_{\mathscr{B}_2}X \in L_{\mathscr{B}_2}^p \subset L^p$, and so $E_{\mathscr{B}_1}(E_{\mathscr{B}_2}X) \in L_{\mathscr{B}_1}^p$. Let q be conjugate to p. Then for every $Y \in L_{\mathscr{B}_1}^q$ we have $Y \in L_{\mathscr{B}_1}^q \subset L_{\mathscr{B}_2}^q$ by $\mathscr{B}_1 \subset \mathscr{B}_2$. Using Theorem 3.8.3 we obtain

$$E((E_{\mathscr{B}_1}(E_{\mathscr{B}_2}X))Y) = E((E_{\mathscr{B}_2}X)Y) = E(XY),$$

and so

$$E_{\mathscr{B}_1}(E_{\mathscr{B}_2}X) = E_{\mathscr{B}_1}X.$$ ∎

Let U be an S-valued random variable. Then $E_{U=u|u=U(\omega)}$ ($=$ the value of the function (of u) $E_{U=u}$ evaluated at $u = U(\omega)$) are denoted by E_U or $E(\cdot|U)$. Since these are equal to $E_{\mathscr{B},\omega}$ for $\mathscr{B} = \bar{\sigma}[U]$ ($\bar{\sigma}[U]$ may be replaced by $\sigma[U]$ if U is a topological random variable), Theorem 3.8.4 ensures that $E_U E_{(U,V)} = E_U$, where (U,V) is the joint variable of U and V.

If $\mathscr{B} = \bar{\sigma}[U]$, then

$$L_{\mathscr{B}}^p = \{\varphi(U) | \varphi \in L^p(S, P^U)\}.$$

Therefore, if $X \in L^2$, then $E_U X$ is the $\varphi(U)$ minimizing

$$\|X - \varphi(U)\|_2, \qquad \varphi \in L^2(S, P^U).$$

Let us observe L^p, $L_{\mathscr{B}}^p$, and $E_{\mathscr{B}}$ from the viewpoint of functional analysis (p and q are assumed to be conjugate):

(i) $L_{\mathscr{B}}^p$ is a closed subspace of L^p.

(ii) $L^q = (L^p)^*$, $L_{\mathscr{B}}^q = (L_{\mathscr{B}}^p)^*$, where the asterisk indicates the conjugate space, the space of all bounded linear functionals and the $=$ should be replaced by $\subset$ in case $p = \infty$.

(iii) $E_{\mathscr{B}}: L^p \to L^p$ is:

(a) linear: $E_{\mathscr{B}}(aX + bY) = aE_{\mathscr{B}}X + bE_{\mathscr{B}}Y$.
(b) positive: $X \geq 0 \Rightarrow E_{\mathscr{B}}X \geq 0$.
(c) bounded: $\|E_{\mathscr{B}}X\|_p \leq \|X\|_p$.
(d) idempotent: $E_{\mathscr{B}}E_{\mathscr{B}} = E_{\mathscr{B}}$.
(e) $\|E_{\mathscr{B}}\| = 1$ ($\|\ \|$: operator norm).

(iv) (Theorem 3.8.4) $(E_{\mathscr{B}}$ on $L_p)^* = E_{\mathscr{B}}$ on L_q ($p \neq \infty$), where the asterisk indicates the conjugate operator.

(v) $\mathscr{B}_1 \subset \mathscr{B}_2 \Rightarrow E_{\mathscr{B}_1} E_{\mathscr{B}_2} = E_{\mathscr{B}_1}$.

(vi) $E_{\mathscr{B}}: L^2 \to L^2$ is the orthogonal projection to $L_{\mathscr{B}}^2$.

Exercise 3.8.

(i) Prove that if $X \in L^p$ is independent of (i.e., $\sigma[X]$ is independent of $\mathscr{B}$), then

$$E_{\mathscr{B}}X = EX.$$

[*Hint:* X is independent of every $Y \in L_{\mathscr{B}}^q$ by the assumption. Use Theorem 3.8.3.]

(ii) Prove that if $\mathscr{B}_1 \supset \mathscr{B}_2 \supset \mathscr{B}_3 \supset \ldots$, $\mathscr{B} = \bigcap_n \mathscr{B}_n$, then

$$\|E_{\mathscr{B}}X - E_{\mathscr{B}_n}X\|_2^2 \to 0, \qquad X \in L^2.$$

[*Hint*: Observe that $L^2_{\mathscr{B}_1} \supset L^2_{\mathscr{B}_2} \supset \cdots$, $L^2_{\mathscr{B}} = \bigcap_n L^2_{\mathscr{B}_n}$. Then the problem is reduced to one on the Hilbert space L^2 by Theorem 3.8.5.]

(iii) Suppose that the probability distribution of (X, Y) is the two-dimensional Gauss distribution

$$N(m, v), \qquad m = (m_1, m_2); \qquad v = (v_{ij})^2_{i, j=1}.$$

Find $E_X Y$. [*Hint*: Find $P_{X=x}(Y \in E)$ first and then integrate to find $E_{X=x}(Y)$. Replacing x by $X(\omega)$, we obtain $E_X Y$. If $\det v > 0$ we can use Exercise 3.5(ii) to find $P_{X=x}(Y \in E)$. If $\det v = 0$, $N(m, v)$ is concentrated on a straight line or a single point.]

4

Sums of independent random variables

The theory of the sum of independent random variables was developed for the purpose of generalizing and elaborating the law of large numbers and was one of the most important fields of probability theory in the 1930s. A number of powerful methods in this theory were introduced by A. Khinchin, A. Kolmogorov, P. Lévy, and others.

4.1   General remarks

We fix a separable perfect probability space (Ω, P) as a base space. Let $X_n(\omega)$ $(n = 1, 2, \ldots)$ be an independent sequence of random variables whose probability distributions are μ_n $(n = 1, 2, \ldots)$, respectively. The joint variable $X(\omega) = (X_1(\omega), X_2(\omega), \ldots)$ is an infinite-dimensional random vector whose probability distribution μ is the complete direct product of μ_n $(n = 1, 2, \ldots)$, that is, $\mu = \overline{\mu_1 \times \mu_2 \times \cdots}$ by Theorem 3.4.5(iii).

Conversely, for every sequence of one-dimensional distributions $\mu_1, \mu_2, \ldots$ we can construct a separable perfect probability space (Ω, P) and a sequence of independent random variables $X_n(\omega)$ $(n = 1, 2, \ldots)$ such that each X_n has probability distribution μ_n. There are a number of ways to achieve such a construction, one of which is as follows.

$$\Omega = \mathbb{R}^\infty, \qquad P = \overline{\mu_1 \times \mu_2 \times \cdots},$$

$$X_n(\omega) = p_n(\omega) \qquad (= \text{the } n\text{th component of } \omega).$$

This is called the *coordinate representation* of $\{X_n\}$.

Let $X_1, X_2, \ldots$ be an independent sequence of random variables and let S_n be the nth partial sum:

$$S_n = X_1 + X_2 + \cdots + X_n, \qquad n = 1, 2, \ldots.$$

The purpose of this chapter is to study the asymptotic behavior of S_n as

165

$n \to \infty$. The event that S_n is convergent is represented by the ω-set

$$C = \left\{ (\forall m)(\exists n)(\forall j, k > n) \, | S_j(\omega) - S_k(\omega) | < 1/m \right\}$$

$$= \bigcap_m \bigcup_n \bigcap_{j > k > n} \left\{ |S_j(\omega) - S_k(\omega)| < 1/m \right\}.$$

Since $|S_j(\omega) - S_k(\omega)|$ is P-measurable in ω, C is a P-measurable set. $P(C)$ is the probability that S_n is convergent.

Theorem 4.1.1. $P(C) = 0$ or 1.

Proof: Let $Y(\omega)$ be a real random variable on (Ω, P) and let $\mathscr{B}(\subset \mathscr{D}(P))$ be a σ-algebra. When $Y(\omega)$ is measurable $\mathscr{B}$, we write $Y \in \mathscr{B}$. It is obvious that

$$Y_1, Y_2, \ldots, Y_n \in \mathscr{B} \Rightarrow Y_1 + Y_2 + \cdots + Y_n \in \mathscr{B}.$$

To prove the theorem we will use Kolmogorov's zero-one law (Theorem 3.4.4). We denote $\sigma[X_1], \sigma[X_2], \ldots$ by $\mathscr{B}_1, \mathscr{B}_2, \ldots$, respectively. Since $X_n \in \mathscr{B}_n$ obviously, we have

$$X_n \in \mathscr{B}_{k\infty} \equiv \bigvee_{j > k} \mathscr{B}_j \qquad (n > k).$$

Hence

$$S_j - S_k = X_{k+1} + X_{k+2} + \cdots + X_j \in \mathscr{B}_{k\infty} \qquad (j > k),$$

$$\left\{ |S_j - S_k| < 1/m \right\} \in \mathscr{B}_{k\infty} \qquad (j > k),$$

$$\bigcap_{j > k > n} \left\{ |S_j - S_k| < 1/m \right\} \in \mathscr{B}_{n\infty}.$$

Since convergence of $S_1, S_2, \ldots$ is equivalent to that of $S_{n+1}, S_{n+2}, \ldots$, C is expressible as follows:

$$C = \bigcap_m \bigcup_{r > n} \bigcap_{j > k > r} \left\{ |S_j - S_k| < 1/m \right\}.$$

Hence

$$C \in \mathscr{B}_{n\infty} \text{ for every } n,$$

and so

$$C \in \bigwedge_n \mathscr{B}_{n\infty} = \bigwedge_n \bigvee_{k > n} \mathscr{B}_k.$$

As $X_1, X_2, \ldots$ are independent, so are $\mathscr{B}_1, \mathscr{B}_2, \ldots$. Therefore, we obtain $P(C) = 0$ or 1 by Kolmogorov's zero-one law. ∎

This theorem claims that for a series of independent real random variables there are two extreme cases, the a.s. convergent case and the a.s.

divergent case and no intermediate case such as the convergence with probability $\frac{1}{2}$. Now the following problems arise:

(1) To find the conditions for convergence;

(2) To find the rate of divergence in the divergent case.

We will discuss these problems in the subsequent sections.

Suppose that $X_1, X_2, \ldots$ are independent, and each X_i is μ_i-distributed (i.e., $P^{X_i} = \mu_i$). Then the probability distribution of $S_n = X_1 + X_2 + \cdots + X_n$ is $\nu_n = \mu_1 * \mu_2 * \cdots * \mu_n$. Hence the characteristic function of S_n is

$$\mathscr{F}\nu_n(z) = \mathscr{F}\mu_1(z)\mathscr{F}\mu_2(z)\ldots\mathscr{F}\mu_n(z).$$

This suggests that the asymptotic behavior of ν_n or $\mathscr{F}\nu_n$ will play an important role to study our problems. It was for this purpose that P. Lévy introduced the notion of characteristic functions of a distribution and a real random variable.

The following inequalities, due to Kolmogorov and Ottaviani, are useful to estimate

$$P\left\{ \max_{k=1}^{n} |X_1 + X_2 + \cdots + X_n| > a \right\}$$

when we are given $\mu_1, \mu_2, \ldots, \mu_n$ (see the remark at the end of this section).

Theorem 4.1.2 (Kolmogorov's inequality). Let $X_1, X_2, \ldots, X_n$ be independent real random variables:

$$EX_i = 0, \qquad v_i \equiv V(X_i) < \infty \qquad (i = 1, 2, \ldots, n).$$

Then

$$P\left\{ \max_{k=1}^{n} |X_1 + X_2 + \cdots + X_k| \geq a \right\} \leq \frac{1}{a^2} \sum_{k=1}^{n} v_k \qquad (a > 0).$$

Proof: Let

$$S_k = X_1 + X_2 + \cdots + X_k,$$

$$T_k = \max_{i=1}^{k} |S_i|,$$

and

$$A_k = \{ T_{k-1} < a, |S_k| \geq a \} \qquad (T_0 = 0).$$

Then $A_1, A_2, \ldots, A_n$ are disjoint and

$$\{ T_n \geq a \} = A_1 + A_2 + \cdots + A_n.$$

Since $X_1, X_2, \ldots$ are independent and $E(X_i) = 0$,

$$E(S_n) = 0 \quad \text{and} \quad V(S_n) = E(S_n^2) = \sum_{k=1}^{n} v_i.$$

Hence the inequality to be proved is written as follows:

$$\sum_{k=1}^{n} P(A_k) \leq \frac{1}{a^2} V(S_n).$$

Let

$$e_k(\omega) = \text{the indicator of } A_k \quad \text{and} \quad \mathscr{B}_k = \sigma[X_k].$$

Since $A_1, A_2, \ldots, A_n$ are disjoint, we have

$$\sum_{k=1}^{n} e_k \leq 1.$$

Hence

(1) $V(S_n) = E(S_n^2) \geq \sum_{k=1}^{n} E(S_n^2 e_k)$,

(2) $E(S_n^2 e_k) = E(S_k^2 e_k) + 2E(S_k(S_n - S_k)e_k) + E((S_n - S_k)^2 e_k)$.

As $X_1, X_2, \ldots, X_n$ are independent, so are $\mathscr{B}_1, \mathscr{B}_2, \ldots, \mathscr{B}_n$. Also,

$$S_k e_k \in \mathscr{B}_1 \vee \mathscr{B}_2 \vee \cdots \vee \mathscr{B}_k,$$

$$S_n - S_k = X_{k+1} + X_{k+2} + \cdots + X_n \in \mathscr{B}_{k+1} \vee \mathscr{B}_{k+2} \vee \cdots \vee \mathscr{B}_n.$$

Therefore $S_k e_k$ and $S_n - S_k$ are independent by Theorem 3.4.3. Since $E(S_n - S_k) = 0$, we have

$$E(S_k(S_n - S_k)e_k) = E(S_k e_k)E(S_n - S_k) = 0.$$

Hence (2) implies

$$E(S_n^2 e_k) \geq E(S_k^2 e_k) = E(S_k^2, A_k),$$

Since $|S_k| \geq a$ on A_k, we obtain

$$E(S_n^2 e_k) \geq a^2 P(A_k),$$

which, combined with (1), yields

$$V(S_n) = E(S_n^2) \geq a^2 \sum_{k=1}^{n} P(A_k). \qquad \blacksquare$$

Theorem 4.1.3 (Ottaviani's inequality). Suppose that $X_1, X_2, \ldots, X_n$ are independent. If

$$P\{|X_{k+1} + X_{k+2} + \cdots + X_n| \leq a\} \geq \beta, \qquad k = 0, 1, 2, \ldots, n-1$$

Then

$$P\left\{\max_{k=1}^{n}|X_1 + X_2 + \cdots + X_k| > 2a\right\} \leq \frac{1}{\beta}P\{|X_1 + X_2 + \cdots + X_n| > a\}.$$

where α, β, and a are positive constants.

Proof: Let S_k, T_k, and $\mathscr{B}_k$ be as before and let

$$A_k = \{T_{k-1} \leq 2a, |S_k| > 2a\},$$

$$B_k = \{|S_n - S_k| \leq a\} \qquad (B_n = \Omega).$$

Then $A_1, A_2, \ldots, A_n$ are disjoint and $\{T_n > 2a\} = A_1 + A_2 + \cdots + A_n$. The inequality to be proved is written as follows:

$$\sum_{k=1}^{n} P(A_k) \leq \frac{1}{\beta}P\{|S_n| > a\}.$$

But $A_1 \cap B_1, A_2 \cap B_2, \ldots, A_n \cap B_n$ are disjoint and

$$\omega \in A_k \cap B_k \Rightarrow |S_k(\omega)| > 2a, \qquad |S_n(\omega) - S_k(\omega)| \leq a$$

$$\Rightarrow |S_n(\omega)| > a$$

Hence

$$\sum_{k=1}^{n} A_k \cap B_k \subset \{|S_n| > a\}.$$

In the same way as in the proof of the last theorem we obtain

$$S_k, T_k \in \mathscr{B}_1 \vee \mathscr{B}_2 \vee \cdots \vee \mathscr{B}_k, \qquad S_n - S_k \in \mathscr{B}_{k+1} \vee \mathscr{B}_{k+2} \vee \cdots \vee \mathscr{B}_n,$$

and so

$$A_k \in \mathscr{B}_1 \vee \mathscr{B}_2 \vee \cdots \vee \mathscr{B}_k, \qquad B_k \in \mathscr{B}_{k+1} \vee \mathscr{B}_{k+2} \vee \cdots \vee \mathscr{B}_n.$$

Hence A_k and B_k are independent by Theorem 3.4.3. and so

$$P(A_k B_k) = P(A_k)P(B_k) \geq P(A_k)\beta$$

This implies

$$P\{|S_n| > a\} \geq \sum_{k=1}^{n} P(A_k \cap B_k) \geq \beta \sum_{k=1}^{n} P(A_k) = \beta P\{T_n > 2a\} \qquad \blacksquare$$

Remark: Let $\mu_i = P^{X_i}$, $i = 1, 2, \ldots, n$. Then the assumptions of Kolmogorov's inequality are written

$$M(\mu_i) = 0, \qquad V(\mu_i) < \infty,$$

and the right-hand side of the equality is

$$\frac{1}{a^2} \sum_{i=1}^{n} V(\mu_i).$$

Also, the assumptions of Ottaviani's inequality are written

$$(\mu_{k+1} * \mu_{k+2} * \cdots * \mu_n)((-a, a)^c) \geq \beta, \qquad k = 0, 1, 2, \ldots (n-1),$$

and the right-hand side of the inequality is

$$\frac{1}{\beta}(\mu_1 * \mu_2 * \cdots * \mu_n)((-a, a)^c).$$

The latter inequality can be used even if $M_2(\mu) = \infty$. The theorems above give probabilistic estimates of $\max_{k=1}^{n} |X_1 + X_2 + \cdots + X_k|$ only in terms of $\mu_1, \mu_2, \ldots, \mu_n$.

Exercise 4.1.

(i) Suppose that μ and ν are one-dimensional distributions and $M_2(\mu)$, $M_2(\nu) < \infty$. Prove that

$$M(\mu * \nu) = M(\mu) + M(\nu), \qquad V(\mu * \nu) = V(\mu) + V(\nu).$$

[*Hint*: Consider two independent random variables X and Y with $P^X = \mu$ and $P^Y = \nu$. Then

$$P^{X+Y} = \mu * \nu,$$

and so

$$M(\mu * \nu) = E(X + Y), \qquad V(\mu * \nu) = V(X + Y).]$$

(ii) Prove that Ottaviani's inequality holds for random vectors or, more generally, for topological random variables with values in a metric vector space.

4.2 Convergent series of independent random variables

Let $X_1, X_2, \ldots$ be independent real random variables. Then the series $\sum_n X_n$ is convergent a.s. or divergent a.s. Now we will find the conditions under which the first case occurs.

Theorem 4.2.1 (Kolmogorov's theorem). If both $\sum_n E(X_n)$ and $\sum_n V(X_n)$ are convergent, $\sum_n X_n$ is convergent a.s.

Proof: Since

$$X_n = (X_n - E(X_n)) + E(X_n), \qquad V(X_n) = V(X_n - E(X_n)),$$

we can assume without any loss of generality that $EX_n = 0$. Let $S_n = X_1 + X_2 + \cdots + X_n$. Applying Kolmogorov's inequality to $X_{n+1}, X_{n+2}, \ldots, X_{n+m}$, we obtain

$$P\left\{ \max_{k=1}^{m} |S_{n+k} - S_n| > \varepsilon \right\} \leq \frac{1}{\varepsilon^2} \sum_{k=1}^{m} V(X_{n+k}).$$

Since

$$|S_{n+k} - S_{n+l}| \leq |S_{n+k} - S_n| + |S_{n+l} - S_n|,$$

we have

$$\max_{1 \leq k, l \leq m} |S_{n+k} - S_{n+l}| \leq 2 \max_{k=1}^{m} |S_{n+k} - S_n|.$$

Hence we obtain

$$P\left\{ \max_{1 \leq k, l \leq m} |S_{n+k} - S_{n+l}| > 2\varepsilon \right\} \leq \frac{1}{\varepsilon^2} \sum_{k=1}^{m} v_{n+k} \qquad (v_i = V(X_i)).$$

As $m \to \infty$, the ω-set in $\{\ \}$ increases and converges to

$$\left\{ \sup_{k,l} |S_{n+k} - S_{n+l}| > 2\varepsilon \right\}.$$

Hence

$$P\left\{ \sup_{k,l} |S_{n+k} - S_{n+l}| > 2\varepsilon \right\} \leq \frac{1}{\varepsilon^2} \sum_{k=1}^{\infty} v_{n+k}.$$

The supremum in $\{\ \}$ decreases as n increases, and so

$$P\left\{ \lim_{n \to \infty} \sup_{k,l} |S_{n+k} - S_{n+l}| > 2\varepsilon \right\} \leq \frac{1}{\varepsilon^2} \sum_{k=1}^{\infty} v_{n+k}, \qquad n = 1, 2, \ldots.$$

Since $\Sigma v_n < \infty$, the right-hand side converges to 0 as $n \to \infty$, and so

$$P\left\{ \lim_{n \to \infty} \sup_{k,l} |S_{n+k} - S_{n+l}| > 2\varepsilon \right\} = 0.$$

Letting $\varepsilon \downarrow 0$, we obtain

$$\lim_{n \to \infty} \sup_{k,l} |S_{n+k} - S_{n+l}| = 0 \text{ a.s.} \qquad \blacksquare$$

This theorem is powerful but it can be used only when every X_n has a finite second-order moment. For the general case, we have the following.

Theorem 4.2.2 (Khinchin's three series theorem). Let

$$X_n' = X_n 1_{[-1,1]}(X_n).$$

ΣX_n is convergent a.s. if and only if the following three series are convergent:

$$\sum_n E(X_n'), \qquad \sum_n V(X_n'), \qquad \sum_n P(X_n \neq X_n').$$

Proof: First note that the independence of $\{X_n'\}$ follows from that of X_n. Suppose that these three series are convergent. By the convergence of the third series we can use the Borel–Cantelli lemma to conclude that

$$P\{X_n = X_n' \text{ f.e.}\} = 1.$$

Hence the almost sure convergence of ΣX_n follows from that of $\Sigma X_n'$. But $\Sigma X_n'$ is convergent a.s. by the convergence of the first two series due to Kolmogorov's theorem.

Suppose conversely that ΣX_n is convergent a.s. Since $A_n \equiv \{X_n \neq X_n'\} \in \sigma[X_n]$, $\{A_n\}$ is independent. If $\Sigma P(A_n)$ is divergent, then Borel's theorem ensures that

$$P\{X_n \neq X_n' \quad \text{i.o.}\} = 1; \text{ that is, } P\{|X_n| > 1 \quad \text{i.o.}\} = 1.$$

Then ΣX_n is divergent a.s. This contradicts the assumption. Hence $\Sigma P(A_n) < \infty$; that is, the third series is convergent. Then

$$P\{X_n = X_n' \text{ f.e.}\} = 1$$

by the Borel–Cantelli lemma. As ΣX_n is convergent a.s. by the assumption, so is $S \equiv \Sigma X_n'$. Denote the characteristic functions of

$$X_n', S_n \equiv \sum_{k=1}^{n} X_k', S$$

by $\varphi_n(z), \psi_n(z), \psi(z)$, respectively, and the probability distribution of X_n' by μ_n. Since $S_n \to S$ a.s. and $\{X_n'\}$ is independent, we obtain

$$|\psi(z)|^2 = \lim_{n \to \infty} |\psi_n(z)|^2 = \lim_{n \to \infty} \prod_{k=1}^{n} |\varphi_k(z)|^2 = \prod_{k=1}^{\infty} |\varphi_k(z)|^2.$$

Since $\psi(z)$ is the characteristic function of S, it does not vanish in a neighborhood U of $z = 0$. Hence

$$\sum_n \left(1 - |\varphi_n(z)|^2\right) < \infty, \qquad z \in U.$$

Let $\nu_n = \mu_n * \breve{\mu}_n$ ($\breve{\mu}_n$: the reflection of μ_n, from Section 2.6.b). Then ν_n is symmetric and concentrated in $[-2, 2]$ and $|\varphi_n(z)|^2 = \mathcal{F}\nu_n$. Hence the inequality above is written as follows:

$$\sum_n \int_{-2}^{2} (1 - \cos z\xi)\nu_n(d\xi) < \infty, \qquad x \in U.$$

If ξ is sufficiently small, then $1 - \cos \xi > \xi^2/3$, so we can find a neighborhood $V = V(0) \subset U$ such that

$$\sum_n \int_{-2}^2 z^2 \xi^2 \nu_n(d\xi) < \infty, \qquad z \in V.$$

Hence

$$\sum_n \int_{\mathbb{R}^1} \xi^2 \nu_n(d\xi) < \infty; \qquad \left(\text{i.e., } \sum_n V(\nu_n) < \infty\right).$$

This implies that

$$\sum_n V(X_n') = \sum_n V(\mu_n) < \infty$$

because

$$V(\nu_n) = V(\mu_n) + V(\check{\mu}_n) = 2V(\mu_n) \qquad (\text{Exercise 4.1(i)}).$$

Therefore, $\Sigma(X_n' - E(X_n'))$ is convergent a.s. Since we have proved above that $\Sigma X_n'$ is convergent a.s., $\Sigma E(X_n')$ must be convergent. ∎

In general, "convergence in distributions," "convergence in probability," and "almost sure convergence" are different from one another (Section 3.7). But the following theorem shows that they are equivalent to each other for series of independent real random variables.

Theorem 4.2.3 (Lévy's equivalence theorem). Let $X_1, X_2, \ldots$ be independent real random variables. Then the following three conditions are equivalent.

  (i) ΣX_n is convergent in distributions;
  (ii) ΣX_n is convergent in probability;
  (iii) ΣX_n is convergent a.s.

Proof: Let

$$S_{mn} = \sum_{k=m+1}^n X_k \qquad (m < n),$$

$$\mu_{m,n} = P^{S_{m,n}},$$

and

$$\varphi_{mn} = \mathscr{F} \mu_{mn}.$$

Since (iii) $\Rightarrow$ (ii) $\Rightarrow$ (i) is obvious, it suffices to prove (i) $\Rightarrow$ (ii) and (ii) $\Rightarrow$ (iii).

Suppose that (i) holds. Then $\mu_{1,n}$ converges to a distribution, say μ, and so $\varphi_{1,n}$ converges to $\varphi = \mathscr{F}\mu$ uniformly on every bounded set. Since φ does not vanish near 0, we can find $a > 0$, $b > 0$, and $N(\varepsilon)$ such that

$$n > N(\varepsilon) \Rightarrow |\varphi_{1n}(z)| \geq \frac{b}{2} \qquad (|z| \leq a)$$

and

$$m > n > N(\varepsilon) \Rightarrow |\varphi_{1n}(z) - \varphi_{1m}(z)| < \frac{b}{2}\varepsilon \qquad (|z| \leq a).$$

Since

$$\varphi_{1m} = \varphi_{1n}\varphi_{nm} \qquad (m > n),$$

we obtain

$$m > n > N(\varepsilon)$$

$$\Rightarrow \left| \int (1 - e^{iz\xi})\mu_{nm}(d\xi) \right| = |1 - \varphi_{nm}(z)| < \varepsilon \qquad (|z| \leq a)$$

$$\Rightarrow \left| \frac{1}{2a} \int_{-a}^{a} dz \int (1 - e^{iz\xi})\mu_{nm}(d\xi) \right| < \varepsilon$$

$$\Rightarrow \int \left(1 - \frac{\sin a\xi}{a\xi} \right)\mu_{nm}(d\xi) < \varepsilon.$$

Taking β sufficiently small so that the integrand is no less than $\beta a^2 \xi^2 (1 + a^2\xi^2)^{-1}$, we have

$$\int \frac{a^2\xi^2}{1 + a^2\xi^2}\mu_{nm}(d\xi) < \varepsilon/\beta, \qquad (m > n > N(\varepsilon)).$$

Hence

$$\mu_{nm}([-\eta, \eta]^c) < \frac{\varepsilon}{\beta} \cdot \frac{1 + a^2\eta^2}{a^2\eta^2} \qquad (\eta > 0, m > n > N(\varepsilon));$$

that is,

$$m, n > N(\varepsilon) \Rightarrow P\{|S_m - S_n| > \eta\} < \varepsilon(1 + a^2\eta^2)\beta^{-1}a^{-2}\eta^{-2} \qquad (\eta > 0).$$

This implies that S_n converges i.p. Thus (i) $\Rightarrow$ (ii) is proved.

Suppose that (ii) holds. Then there exists $N(\varepsilon)$ such that

$$P\{|S_m - S_n| > \varepsilon\} < \tfrac{1}{2} \qquad (m > n > N(\varepsilon)),$$

which, combined with Ottaviani's inequality, implies that

$$P\left\{ \max_{k=1}^{m} |S_{n+k} - S_n| > 2\varepsilon \right\} \leq 2P\{|S_{n+m} - S_n| > \varepsilon\}.$$

Using the same argument as in the last step of the proof of Theorem 4.2.1, we can prove that S_n is convergent a.s. Thus (ii) → (iii) is proved.

∎

Exercise 4.2.

(i) Let $X_1, X_2, \ldots$ be independent real random variables such that P^{X_n} is the uniform distribution on $[-1/n, 1/n]$. Prove that (a) ΣX_n is convergent a.s. and (b) $\Sigma |X_n|$ is divergent a.s. (Hence ΣX_n is conditionally convergent a.s.) [*Hint*: Use Kolmogorov's theorem for (a) and Khinchin's theorem for (b).]

(ii) Let $X_1, X_2, \ldots$ be independent real random variables such that P^{X_n} is $N(m_n, v_n)$. Prove that ΣX_n is convergent a.s. if and only if both Σm_n and Σv_n are convergent. [*Hint*: The "if" part follows from Kolmogorov's theorem. Suppose that

$$S_n = \sum_{k=1}^{n} X_k \to S \text{ a.s.}$$

By the independence of X_n we have

$$P^{S_n} = N(M_n, V_n) \qquad \left(M_n = \sum_{1}^{n} m_k, \ V_n = \sum_{1}^{n} v_k\right)$$

and so

$$|\mathscr{F}P^S(z)| = \lim_{n \to \infty} |\mathscr{F}P^{S_n}(z)| = \lim_{n \to \infty} e^{-V_n z^2/2}.$$

Since the left-hand side does not vanish near $z = 0$, $\{V_n\}$ should be bounded; that is, Σv_n is convergent. Hence $\Sigma(X_n - m_n)$ is convergent a.s. by Kolmogorov's theorem. But ΣX_n is convergent a.s. by the assumption. Hence Σm_n must be convergent.]

4.3 Central values and dispersions

Let $\{X_k\}$ be an independent sequence of square integrable real random variables with the partial sums

$$S_n = X_1 + X_2 + \cdots + X_n, \qquad n = 1, 2, \ldots .$$

If both $\{E(S_n)\}$ and $\{V(S_n)\}$ are convergent, then $\{S_n\}$ is convergent a.s. The converse is also true if S_n is Gauss-distributed; that is, if P^{S_n} is a Gauss distribution (Exercise 4.2(ii)). The purpose of this section is to establish a similar fact without any assumption of integrability of X_n. For this purpose we introduce the central value and the dispersion of a general real random variable that play the same roles as the mean value and the variance of a square integrable random variable, respectively.

Let X be a real random variable. Since arctan is a homeomorphic map from $(-\infty, \infty)$ to $(-\pi/2, \pi/2)$, there exists a unique real number γ such that $E(\arctan(X - \gamma)) = 0$. This number was introduced by J. L. Doob and is called the *central value* of X, written $\gamma(X)$. It is obvious by the definition that

$$\gamma(X + a) = \gamma(X) + a \qquad (a\colon \text{real constant}),$$
$$\gamma(-X) = -\gamma(X).$$

Also, the *dispersion* of X, $\delta(X)$ is defined as

$$\delta(X) = -\log\left[\iint e^{-|\xi - \eta|}\mu(d\xi)\mu(d\eta)\right] \in [0, \infty) \qquad (\mu = P^X).$$

It is obvious that

$$\delta(X + a) = \delta(X), \qquad \delta(-X) = \delta(X).$$

The mean value $E(X)$ $(X \in L^1)$ is defined to be the real number m such that $E(X - m) = 0$. Hence the central value is similar to the mean value and can be defined for every random variable.

The variance $V(X)$ $(X \in L^2)$ is given by

$$V(X) = \inf_{-\infty < m < \infty} E((X - m)^2),$$

because

$$E((X - m)^2) = V(X) + (m - EX)^2.$$

Hence $V(X)$ indicates the degree of scattering of values of X. $\delta(X)$ $(X \in L^0)$ is a quantity of the same character, as the following facts shows:

$$\mu[a - l, a + l]^2 = \iint_{|\xi - a|, |\eta - a| \leq l} \mu(d\xi)\mu(d\eta)$$

$$\leq \iint_{|\xi - \eta| \leq 2l} \mu(d\xi)\mu(d\eta),$$

$$e^{-\delta(X)} = \iint e^{-|\xi - \eta|}\mu(d\xi)\mu(d\eta) \geq e^{-2l}\iint_{|\xi - \eta| \leq 2l} \mu(d\xi)\mu(d\eta),$$

$$\mu[a - l, a + l] \leq e^{l - \delta(X)/2},$$

and so if $\delta(X)$ is very large, then the probability that X lies in an interval of length $2l$ is very small. Also, we have

$$e^{-\delta(X)} = \iint e^{-|\xi-\eta|}\mu(d\xi)\mu(d\eta)$$

$$\leq \int \left(e^{-l}\mu([-l+\eta, l+\eta]^c) + \mu[-l+\eta, l+\eta] \right)\mu(d\eta)$$

$$\leq e^{-l} + \sup_a \mu[a-l, a+l],$$

and so $\delta(X)$ is very large if the probability that X lies in a very large interval is very small.

It holds that

$$\delta(X) = 0 \Leftrightarrow X = a \text{ a.s.} \qquad (a: \text{constant}).$$

In fact, the implication ($\Leftarrow$) is obvious, whereas $\delta(X) = 0$ implies

$$\iint e^{-|\xi-\eta|}\mu(d\xi)\mu(d\eta) = 1 \qquad (\mu = P^X),$$

and so

$$\int e^{-|\xi-a|}\mu(d\xi) = 1$$

for a constant a, and this implies

$$\mu\{a\} = 1 \qquad \text{i.e., } X = a \text{ a.s.}$$

It also holds that

$$\delta(X_n) \to 0 \Leftrightarrow \exists \{a_n\}: X_n - a_n \to 0 \text{ i.p.}$$

Suppose that $\delta(X_n) \to 0$. Then

$$\iint e^{-|\xi-\eta|}\mu_n(d\xi)\mu_n(d\eta) \to 1 \qquad (\mu_n = P^{X_n}),$$

and so

$$\sup_\eta \int e^{-|\xi-\eta|}\mu_n(d\xi) \to 1.$$

Hence we can find $\{a_n\}$ such that

$$\int e^{-|\xi-a_n|}\mu_n(d\xi) \to 1, \qquad E(e^{-|X_n-a_n|}) \to 1.$$

Using Bienaymé's inequality, we obtain

$$P(|X_n - a_n| > \varepsilon) \leq \frac{1}{1-e^{-\varepsilon}}E(1 - e^{-|X_n-a_n|}) \to 0;$$

that is,

$$X_n - a_n \to 0 \text{ i.p.}$$

Suppose conversely that $X_n - a_n \to 0$ i.p. Then $\nu_n = P^{X_n - a_n} \to \delta$ and so $\nu_n \times \nu_n \to \delta \times \delta$. Since $e^{-|\xi - \eta|}$ is bounded and continuous in (ξ, η), we obtain

$$e^{-\delta(X_n)} = e^{-\delta(X_n - a_n)} = \iint e^{-|\xi - \eta|} \nu_n(d\xi) \nu_n(d\eta) \to 1;$$

that is, $\delta(X_n) \to 0$. $\blacksquare$

If $\{X_n\} \subset L^2$ is independent, then the variance of $S_n = \Sigma_{k=1}^n X_k$ increases with n. The dispersion δ has the same property, as the following theorem shows.

Theorem 4.3.1. Let $\{X_n\} \subset L^0$ be independent. Then the dispersion of $S_n = \Sigma_1^n X_k$ increases with n.

Proof: First observe that

$$\iint e^{-|\xi - \eta|} \mu(d\xi) \mu(d\eta) = \int e^{-|\xi|} (\mu * \check{\mu})(d\xi) \qquad (\check{\mu}: \text{ the reflection of } \mu)$$

$$= \frac{1}{\pi} \iint \frac{e^{i\xi x}}{1 + x^2} dx (\mu * \check{\mu})(d\xi)$$

$$= \frac{1}{\pi} \int \frac{dx}{1 + x^2} \int e^{i\xi x} (\mu * \check{\mu})(d\xi)$$

$$= \frac{1}{\pi} \int \frac{dx}{1 + x^2} |\mathscr{F}\mu(x)|^2.$$

Let φ_n be the characteristic function of X_n. Then the characteristic function of S_n is $\Pi_1^n \varphi_k$. Hence

$$e^{-\delta(S_n)} = \frac{1}{\pi} \int \frac{dx}{1 + x^2} \left| \prod_{k=1}^n \varphi_k(x) \right|^2.$$

Since $|\varphi_k(x)| \leq 1$, the right-hand side decreases and so $\delta(S_n)$ increases as n increases. $\blacksquare$

In case $\{X_n\} \subset L^2$, $V(S_n) = V(S_{n+1})$ implies that $V(X_{n+1}) = 0$; that is, X_{n+1} equals a constant a.s. We will prove the corresponding fact for $\{X_n\} \subset L^0$.

Suppose that $\delta(S_n) = \delta(S_{n+1})$. Then

$$\left|\prod_1^n \varphi_k(x)\right|^2 = \left|\prod_1^{n+1} \varphi_k(x)\right|^2 \quad \text{a.e. on } R.$$

Since $\varphi_k(x)$ is continuous and equals 1 at $x = 0$, φ_k is near 1 on an interval $-a \leq x \leq a$ for $k = 1, 2, \ldots, n$ and so

$$|\varphi_{n+1}(x)|^2 = 1 \quad (|x| \leq a),$$

because $|\varphi_{n+1}(x)| \leq 1$ always. Let μ denote the probability distribution of X_{n+1}. Then $|\varphi_{n+1}|^2$ is equal to $\mathscr{F}\tilde{\mu}$ ($\tilde{\mu} = \mu * \check{\mu}$). Since $\tilde{\mu}$ is symmetric, we have

$$\int \cos x\xi \, \tilde{\mu}(d\xi) = 1;$$

that is,

$$\int (1 - \cos x\xi) \tilde{\mu}(d\xi) = 0.$$

Taking the average on $[-a, a]$, we have

$$\int \left(1 - \frac{\sin a\xi}{a\xi}\right) \tilde{\mu}(d\xi) = 0 \qquad \left(\frac{\sin 0}{0} = 1 \text{ by convention}\right).$$

Since the integrand is positive for $\xi \neq 0$, $\tilde{\mu}$ must be concentrated at 0. Hence μ is concentrated at a constant; that is, X_{n+1} is a constant a.s.

The following theorem answers problem (1) raised in Section 4.1.

Theorem 4.3.2. Let $\{X_n\}$ be an independent sequence of random variables and let

$$S_n = X_1 + X_2 + \cdots + X_n, \qquad n = 1, 2, \ldots.$$

Then
   (i) $\exists a_n$: $\{S_n - a_n\}$ is convergent a.s. if and only if $\{\delta(S_n)\}$ is convergent
   (ii) $\{S_n\}$ is convergent a.s. if and only if both $\{\delta(S_n)\}$ and $\{\gamma(S_n)\}$ are convergent.

Proof: Since $\delta(S_n)$ increases with n, $\delta_\infty = \lim_{n \to \infty} \delta(S_n) \in [0, \infty]$ is determined.

(i) Suppose that $T_n = S_n - a_n \to T$ a.s. As we observed in the proof of the last theorem, we have

$$\delta(S_n) = \delta(T_n) - \log\left(\frac{1}{\pi} \int \frac{dx}{1+x^2} |E(e^{izT_n})|^2\right).$$

Hence we obtain

$$\delta_\infty = -\log\left(\frac{1}{\pi} \int \frac{dx}{1+x^2} |E(e^{izT})|^2\right) = \delta(T) < \infty$$

by the bounded convergence theorem.

Suppose conversely that $\delta_\infty < \infty$. Let μ_n and φ_n denote the probability distribution and the characteristic function of X_n, respectively. Then

$$\delta_\infty = \lim_{n \to \infty} \delta(S_n) = -\log\left(\frac{1}{\pi} \int \frac{dx}{1+x^2} \prod_{n=1}^{\infty} |\varphi_n(x)|^2\right).$$

Since $\delta_\infty < \infty$,

$$\int \frac{dx}{1+x^2} \prod_{n=1}^{\infty} |\varphi_n(x)|^2 > 0.$$

Hence

$$\prod_{n=1}^{\infty} |\varphi_n(x)|^2 > 0,$$

on a set A of positive Lebesgue measure, that is,

$$\sum_{n=1}^{\infty} \left(1 - |\varphi_n(x)|^2\right) < \infty \qquad (x \in A).$$

Hence we can find a positive number C and a bounded set $B(\subset A)$ of positive Lebesgue measure such that

$$\sum_n \left(1 - |\varphi_n(x)|^2\right) < C \qquad (x \in B).$$

We can assume that the Lebesgue measure of B is no more than 1. Since $|\varphi_n(x)|^2$ is the characteristic function of $\tilde{\mu}_n = \mu_n * \check{\mu}_n$, the inequality implies that

$$\sum_n \int (1 - \cos x\xi)\tilde{\mu}_n(d\xi) < C \qquad (x \in B),$$

so

$$\sum_n \int\int_B (1 - \cos x\xi) \, dx \, \tilde{\mu}_n(d\xi) < C.$$

As we will prove later, it holds that

$$(*) \qquad \int_B (1 - \cos x\xi)\, dx \geq C_1 \frac{\xi^2}{1 + \xi^2} \qquad (C_1: \text{positive constant}).$$

Hence,

$$\sum_n \int \frac{\xi^2}{1 + \xi^2} \tilde{\mu}_n(d\xi) < C_2 \equiv \frac{C}{C_1};$$

that is,

$$\sum_n \iint \frac{(\xi - \eta)^2}{1 + (\xi - \eta)^2} \mu_n(d\xi)\mu_n(d\eta) < C_2,$$

and so

$$\sum_n \inf_\eta \int \frac{(\xi - \eta)^2}{1 + (\xi - \eta)^2} \mu_n(d\xi) < C_2.$$

Hence we can find a sequence $\{c_n\}$ such that

$$\sum_n \int \frac{(\xi - c_n)^2}{1 + (\xi - c_n)^2} \mu_n(d\xi) < C_2.$$

This implies that

$$\sum_n \int_{|\xi - c_n| \leq 1} (\xi - c_n)^2 \mu_n(d\xi) < \infty, \qquad \sum_n \int_{|\xi - c_n| > 1} \mu_n(d\xi) < \infty.$$

Define Y_n to be $X_n - c_n$ or 0 according to whether $|X_n - c_n| \leq 1$ or not. Then these inequalities imply that

$$\sum_n V(Y_n) \leq \sum_n E(Y_n^2) < \infty, \qquad \sum_n P(Y_n \neq X_n - c_n) < \infty.$$

The first inequality, together with Kolmogorov's theorem (Theorem 4.2.1), implies that $\sum_n (Y_n - E(Y_n))$ is convergent a.s. The second inequality, together with the Borel–Cantelli lemma, implies that $Y_n = X_n - c_n$ (f.e.) a.s. Therefore $\sum (X_n - c_n - E(Y_n))$ is convergent a.s.; namely,

$$S_n - \sum_{k=1}^n (c_k + E(Y_k)), \qquad n = 1, 2, \ldots$$

is convergent a.s.

To complete the proof of (i) it remains only to prove the inequality $(*)$. Consider the function of $\xi \in \mathbb{R} - \{0\}$:

$$f(\xi) = \frac{1 + \xi^2}{\xi^2} \int_B (1 - \cos x\xi)\, dx = \frac{1 + \xi^2}{\xi^2}\left(|B| - \int \cos x\xi 1_B(x)\, dx\right),$$

where $|B|$ is the Lebesgue measure of B. Since $|B| > 0$ and $1 - \cos x\xi > 0$ except on a countable x-set for every $\xi \in \mathbb{R}$, $f(\xi)$ is positive for every $\xi \in \mathbb{R} - \{0\}$. Since the integral on the right-hand side converges to 0 as $|\xi| \to \infty$ by the Riemann–Lebesgue theorem, we have

$$\lim_{|\xi| \to \infty} f(\xi) = |B| \in (0, \infty).$$

If $\xi \to 0$, then

$$f(\xi) = \int_B (1 - \cos x\xi) \frac{1 + \xi^2}{\xi^2} dx \to \int_B \frac{1}{2} x^2 dx \in (0, \infty)$$

by the bounded convergence theorem. It is easy to check that $f(\xi)$ is continuous in $\xi \in \mathbb{R} - \{0\}$. Hence

$$C_1 = \inf_\xi f(\xi) > 0.$$

The inequality $(*)$ holds with this positive number C_1.

(ii) Suppose that S_n converges to S a.s. Then $\{\delta(S_n)\}$ is convergent by (i). We will prove that $\gamma(S_n) \to \gamma(S)$. It suffices to prove that if

$$\gamma(S_{p(n)}) \to l \in [-\infty, \infty],$$

then $l = \gamma(S)$. By the assumption we have

$$\arctan(S_{p(n)} - \gamma(S_{p(n)})) \to \arctan(S - l) \text{ a.s.}$$

$$(\arctan(\pm\infty) = \pm\pi/2).$$

Hence $E(\arctan(S - l)) = 0$, by the bounded convergence theorem and the definition of γ. Hence $l = \gamma(S)$.

Suppose conversely that both $\{\gamma(S_n)\}$ and $\{\delta(S_n)\}$ are convergent. Since $\{\delta(S_n)\}$ is convergent, we can use (i) to find a sequence $\{a_n\}$ such that $\{S_n - a_n\}$ is convergent a.s. Hence $\{\gamma(S_n - a_n)\}$ is convergent as we have proved above. $\{\gamma(S_n)\}$ is convergent by the assumption. Hence

$$\{a_n \equiv \gamma(S_n) - \gamma(S_n - a_n)\}$$

is also convergent. As $\{S_n - a_n\}$ is convergent a.s., so is $\{S_n\}$. ∎

Remark: In the proof of (ii) we have shown that

$$S_n \to S \text{ a.s.} \Rightarrow \gamma(S_n) \to \gamma(S)$$

for every sequence of real random variables. Using this fact we can see that statement (i) holds for $a_n = \gamma(S_n)$. In fact, if $\{S_n - a_n\}$ is convergent a.s. for a sequence $\{a_n\}$, then

$$\gamma(S_n) - a_n = \gamma(S_n - a_n), \qquad n = 1, 2, \ldots,$$

is convergent and so

$$S_n - \gamma(S_n) = (S_n - a_n) - (\gamma(S_n) - a_n), \qquad n = 1, 2, \ldots,$$

is convergent.

Exercise 4.3.

(i) Let $\{X_k\} \subset L^2$ be an independent sequence of real random variables such that $\sum_n V(X_n) < \infty$. Prove that $\sum_n X_n$ is convergent a.s. if and only if $\sum_n E(X_n)$ is convergent. [*Hint:* $\sum_n (X_n - EX_n)$ is convergent by Theorem 4.2.1.]

(ii) Prove that if $X_n \to X$ i.p. then $\gamma(X_n) \to \gamma(X)$ and $\delta(X_n) \to \delta(X)$. (Note that the independence of $\{X_n\}$ is not assumed.) [*Hint:* If $\gamma(X_{p(n)}) \to l$, then a subsequence of $\{X_{p(n)}\}$, say, $\{X_{q(n)}\}$, converges a.s. to X by the assumption. Hence $\gamma(X_{q(n)}) \to \gamma(X)$ and so $l = \gamma(X)$. Also,

$$e^{-\delta(X_n)} = \frac{1}{\pi} \int \frac{dz}{1+z^2} |E(e^{izX_n})|^2$$

$$\to \frac{1}{\pi} \int \frac{dz}{1+z^2} |E(e^{izX})|^2 = e^{-\delta(X)}.]$$

(iii) For a one-dimensional distribution μ we define its central value $\gamma(\mu)$ to be the γ satisfying $\int \arctan(\xi - \gamma)\mu(d\xi) = 0$ and its dispersion $\delta(\mu)$ by $\delta(\mu) = -\log(\int\int e^{-|\xi - \eta|}\mu(d\xi)\mu(d\eta))$. Prove that if $\mu_n \to \mu$, then $\gamma(\mu_n) \to \gamma(\mu)$ and $\delta(\mu_n) \to \delta(\mu)$. [*Hint:* To prove $\gamma(\mu_n) \to \gamma(\mu)$, show that any limiting point of $\{\gamma(\mu_n)\}$ is equal to $\gamma(\mu)$, observing that $|\arctan x - \arctan y| \le |x - y|$. To prove $\delta(\mu_n) \to \delta(\mu)$, observe that the double integral above is equal to

$$\frac{1}{\pi} \int \frac{dx}{1+x^2} |(\mathscr{F}\mu)(x)|^2.]$$

(iv) Use (iii) to prove that if the conditions of Theorem 2.5.3 are satisfied, then both $\delta(\mu)$ and $\gamma(\mu)$ $(\mu \in \mathscr{M})$ are bounded. [*Hint:* Suppose that $\delta(\mu_n) \to \infty$ for a sequence $\mu_n \in \mathscr{M}$. $\{\mu_n\}$ has a convergent subsequence $\{\mu_{p(n)}\}$ for which $\delta(\mu_{p(n)}) \to \infty$, in contradiction with (iii). Similarly for $\gamma(\mu)$.]

4.4 Divergent series of independent random variables

Let $\{X_n\}$ be an independent sequence of real random variables and let

$$S_n = \sum_{k=1}^{n} X_k, \qquad n = 1, 2, \ldots .$$

In the last section we have proved that the following three cases exhaust all possibilities:

(C) Both $\{\delta(S_n)\}$ and $\{\gamma(S_n)\}$ are convergent. In this case $\{S_n\}$ is convergent a.s.

(D_0) $\{\delta(S_n)\}$ is convergent, but $\{\gamma(S_n)\}$ is divergent. In this case $\{S_n\}$ is divergent a.s. but $\{S_n - \gamma(S_n)\}$ is convergent a.s.

(D) $\{\delta(S_n)\}$ is divergent. In this case $\{S_n - a_n\}$ is divergent a.s. for every sequence $\{a_n\}$.

ΣX_n is said to be of the *convergent type* in case (C), of the *pseudodivergent type* in case (D_0), and of the *properly divergent type* in case (D). Here "divergent" means "not convergent" and so includes "oscillating."

In the pseudodivergent case (D_0) $\{S_n - \gamma(S_n)\}$ is convergent a.s. and so the order of divergence of

$$S_n = (S_n - \gamma(S_n)) + \gamma(S_n), \qquad n = 1, 2, \ldots,$$

is determined by that of $\{\gamma(S_n)\}$.

Now we observe the properly divergent case. Since $\delta(S_n) \to \infty$ $(n \to \infty)$, we can choose a subsequence $\{S_{p(n)}\}$ such that

$$\sum_{n=1}^{\infty} e^{-\delta(S_{p(n)})/2} < \infty \qquad \left(\text{e.g., } \delta(S_{p(n)}) > n\right).$$

Let $\{a_n\}$ be an arbitrary sequence of real numbers. Since

$$P\{|S_n - a_n| \leq l\} \leq e^{l - \delta(S_n)/2} \qquad \text{(see the last section),}$$

we obtain

$$\sum_n P\{|S_{p(n)} - a_{p(n)}| \leq l\} \leq e^l \sum_n e^{-\delta(S_{p(n)})/2} < \infty .$$

By the Borel–Cantelli lemma this implies that

$$P\{|S_{p(n)} - a_{p(n)}| > l \text{ f.e.}\} = 1.$$

Hence

$$P\left\{\limsup_{n \to \infty} |S_n - a_n| \geq l\right\} = 1.$$

Letting $l \to \infty$ we obtain the following theorem.

Theorem 4.4. In the properly divergent case we have $\limsup_{n \to \infty} |S_n - a_n| = \infty$ a.s. for every sequence $\{a_n\} \subset \mathbb{R}$.

Remark: This theorem implies that the probability that $\{S_n - a_n\}$ is confined in a fixed finite interval I is 0. But the probability that $S_n - a_n$ enters I infinitely often is not necessarily 0. For example, if

$$P\{X_n = -1\} = P\{X_n = 1\} = \tfrac{1}{2}, \qquad n = 1, 2, \ldots,$$

then

$$E(e^{izX_n}) = \cos z,$$

$$E(e^{izS_n}) = (\cos z)^n,$$

and

$$e^{-\delta(S_n)} = \frac{1}{\pi} \int \frac{(\cos x)^{2n}}{1 + x^2} dx \to 0 \qquad \left(\text{i.e., } \delta(S_n) \to \infty\right)$$

and so ΣX_n is of the properly divergent type. Therefore $\limsup_{n \to \infty} |S_n| = \infty$ a.s. by the theorem above. Nevertheless,

$$P(S_n = 0 \text{ i.o.}) = 1,$$

as we will prove in Section 4.7.

In case $\delta(S_n)$ increases so rapidly that

$$\sum_n e^{-\delta(S_n)/2} < \infty,$$

we can easily check that

$$\lim_{n \to \infty} |S_n - a_n| = \infty \text{ a.s.}$$

for every sequence $\{a_n\}$.

Exercise 4.4.

(i) Let $\{X_n\}$ be an independent sequence of random variables, each X_n having probability distribution $N(m_n, v_n)$. Express a necessary and sufficient condition for ΣX_n to be of the convergent (or pseudodivergent or properly divergent) type in terms of Σm_n and Σv_n. [*Hint*: Let $S_n = \Sigma_{k=1}^n X_k$. Then

$$\gamma(S_n) = \sum_{1}^{n} m_k,$$

$$e^{-\delta(S_n)} = \frac{1}{\pi} \int \frac{dx}{1 + x^2} \exp\left\{-x^2 \sum_{1}^{n} v_k\right\},$$

$$\lim_n \delta(S_n) < \infty \Leftrightarrow \sum_{1}^{\infty} v_k < \infty.]$$

(ii) Let $\{X_n\}$ be an independent sequence of real random variables with the same probability distribution μ. (In this case we say that X_n, $n = 1, 2, \ldots$, are *independently identically distributed* (μ) or *i.i.d.* (μ).) Prove that if μ is not concentrated at a single point, then ΣX_n is of the properly divergent type. [*Hint*: Let $S_n = \Sigma_1^n X_k$. Then

$$e^{-\delta(S_n)} = \frac{1}{\pi} \int \frac{dx}{1 + x^2} |\varphi(x)|^{2n} \qquad (\varphi = \mathscr{F}\mu).$$

If we prove that $|\varphi(x)| \neq 1$ except for a countable number of values of x, then the integral converges to 0 as $n \to \infty$ and $\delta(S_n) \to \infty$. Suppose that

$$|\varphi(x_1)| = |\varphi(x_2)| = 1, \qquad x_1, x_2 \neq 0; \, x_1 \neq x_2.$$

Then

$$\int \cos x_1 \xi \tilde{\mu}(d\xi) = \int \cos x_2 \xi \tilde{\mu}(d\xi) = 1 \qquad (\tilde{\mu} = \mu * \tilde{\mu}).$$

Hence

Supp $\tilde{\mu} \subset \{2\pi n/x_i, n = 0, \pm 1, \pm 2, \ldots\}$ ($i = 1, 2$; Supp: Support).

If μ is not concentrated at a single point, then the support of $\tilde{\mu}$ contains at least one point $\xi_0 \neq 0$ that can be written $\xi_0 = 2\pi n_i/x_i$ ($i = 1, 2$; $n_1, n_2 \neq 0$). This implies that x_1/x_2 is rational. Hence $|\varphi(x)| = 1$ holds only for a countable number of values of x.]

4.5   **Strong law of large numbers**

Let $\{X_n\}$ be an independent sequence of random variables and let

$$S_n = \sum_{k=1}^{n} X_k, \qquad n = 1, 2, \ldots.$$

In case ΣX_n is of the properly divergent type we have

$$\limsup_{n \to \infty} |S_n - a_n| = \infty \text{ a.s.}$$

for every sequence $\{a_n\} \subset \mathbb{R}$ (Theorem 4.4). But it may happen that for a given sequence of positive numbers $b_n \to \infty$ we can find a sequence $\{a_n\}$ such that

$$\limsup_{n \to \infty} \frac{|S_n - a_n|}{b_n} = 0;$$

that is,

$$\lim_{n \to \infty} \frac{S_n - a_n}{b_n} = 0 \text{ a.s.}$$

In this case $\{|S_n - a_n|\}$ is very small compared with $\{b_n\}$ as $n \to \infty$. Here we will discuss some estimation of this type.

Lemma 4.5 (Kronecker's lemma). Let $\{x_n\}$ be a sequence of real numbers and let $\{b_n\}$ be an increasing sequence of positive numbers such that $b_n \to \infty$. If $\sum_n x_n/b_n$ is convergent, then

$$\frac{1}{b_n} \sum_{k=1}^{n} x_k \to 0 \qquad (n \to \infty).$$

Proof: We set

$$s_n = \sum_{k=1}^{n} \frac{x_k}{b_k}, \qquad n = 1, 2, 3, \ldots,$$

$$s_0 = 0,$$

and

$$s_\infty = \lim_{n \to \infty} s_n.$$

Then

$$x_n = b_n(s_n - s_{n-1}).$$

Using Abel's summation formula, we obtain

$$\frac{1}{b_n} \sum_{k=1}^{n} x_k = \frac{1}{b_n} \sum_{k=1}^{n} b_k(s_k - s_{k-1}) = s_n - \frac{1}{b_n} \sum_{k=1}^{n} s_{k-1}(b_k - b_{k-1})$$

$$(b_0 = 0).$$

Also, we have

$$b_k - b_{k-1} \geq 0, \qquad \sum_{k=1}^{n} (b_k - b_{k-1}) = b_n \to \infty; s_n \to s_\infty.$$

Hence

$$\lim_{n \to \infty} \frac{1}{b_n} \sum_{k=1}^{n} x_k = s_\infty - s_\infty = 0. \qquad \blacksquare$$

Theorem 4.5.1. Suppose that $\{X_n\} \subset L^2$ is an independent sequence, $b_n > 0$, and $b_n \uparrow \infty$. Then

$$\sum_n \frac{V(X_n)}{b_n^2} < \infty \Rightarrow \frac{S_n - E(S_n)}{b_n} \to 0 (n \to \infty) \text{ a.s.}$$

Proof: Observing $X_n - E(X_n)$ instead of X_n we can reduce this theorem to the case where $E(X_n) = 0$ and so $E(S_n) = 0$. The sequence

$$Y_n = X_n/b_n, \qquad n = 1, 2, \ldots,$$

is also independent and satisfies

$$EY_n = 0, \qquad \sum_n V(Y_n) = \sum_n \frac{V(X_n)}{b_n^2} < \infty.$$

Hence Theorem 4.2.1 (Kolmogorov's theorem) ensures that $\sum X_n/b_n$ is convergent a.s. Using Kronecker's lemma we can conclude that $\{S_n/b_n\}$ converges to 0 a.s. ∎

Let $\{X_n\} \subset L^2$ be an independent sequence. If $\sum V(X_n)$ is convergent, $S_n - E(S_n)$ ($S_n = \sum_{k=1}^n X_k$) is convergent a.s. by Kolmogorov's theorem and hence $(S_n - E(S_n))/b_n \to 0$ for every sequence $b_n \to \infty$. If $\sum V(X_n)$ is divergent, we can use the theorem above to obtain the following theorem.

Theorem 4.5.2. If $\{X_n\} \subset L^2$ is an independent sequence and $\sum V(X_n)$ is divergent, then

$$\frac{S_n - E(S_n)}{V(S_n)^{1/2+\varepsilon}} \to 0 \text{ a.s.}$$

for every $\varepsilon > 0$.

Proof: Since $V(S_n) \to \infty$, we can assume without any loss of generality that $V(S_1) > 0$. Let $b_n = V(S_n)^{1/2+\varepsilon}$. Observing that $V(X_n) = V(S_n) - V(S_{n-1})$, we obtain

$$\sum_{n=2}^\infty \frac{V(X_n)}{b_n^2} \leq \sum_{n=2}^\infty \int_{V(S_{n-1})}^{V(S_n)} \frac{dx}{x^{1+2\varepsilon}} = \int_{V(S_1)}^\infty \frac{dx}{x^{1+2\varepsilon}} < \infty,$$

which, combined with the last theorem, completes the proof. ∎

As a special case of this theorem we obtain the following:

Theorem 4.5.3 (Kolmogorov's strong law of large numbers). If $\{X_n\} \subset L^2$ is an independent sequence and $\{V(X_n)\}$ is bounded, then

$$\frac{S_n - E(S_n)}{n} \to 0 \text{ a.s.}$$

Proof: If $\sum V(X_n)$ is convergent, then $S_n - E(S_n)$ is convergent a.s. and our assertion is trivial. If $\sum V(X_n)$ is divergent, we set $\varepsilon = \frac{1}{2}$ in the last theorem to obtain

$$\frac{S_n - E(S_n)}{V(S_n)} \to 0 \text{ a.s.}$$

Since $V(X_n)$ is bounded, we have

$$V(S_n) = \sum_{k=1}^{n} V(X_n) \leqq na \qquad (a = \sup V(X_n) < \infty),$$

and so $(S_n - E(S_n))/n \to 0$ a.s. obviously. ∎

Under the same assumption as above we can use Chebyshev's inequality to check that

$$P\left\{ \left| \frac{S_n - E(S_n)}{n} \right| > \varepsilon \right\} \leqq (n\varepsilon)^{-2} n \sup_n V(X_n),$$

which implies

$$\frac{S_n - E(S_n)}{n} \to 0 \text{ i.p.}$$

This is a generalization of Bernoulli's law of large numbers mentioned in Section 1.7. Compared with this, Kolmogorov's strong law of large numbers gives more precise information.

In the theorem above we assumed that $\{X_n\} \subset L^2$. In case all X_n's have the same probability distribution, we have the same conclusion even if $\{X_n\} \subset L^1$, as we can see in the following theorem.

Theorem 4.5.4 (The strong law of large numbers for identically distributed random variables). If $X_1, X_2, \ldots \in L^1$ are independently identically distributed (μ), then $S_n/n \to M(\mu)$ a.s.

Proof: Set $X_n' = X_n$ or 0 according to whether $|X_n| \leqq n$ or not and set $S_n' = \sum_{k=1}^{n} X_k'$. Let ν denote the probability distribution of $|X_n|$, which is clearly independent of n. Then

$$\sum_{n=1}^{\infty} P\{X_n \neq X_n'\} = \sum_{n=1}^{\infty} P\{|X_n| > n\} = \sum_{n=1}^{\infty} \nu(n, \infty)$$

$$\leqq \int_0^{\infty} \nu(y, \infty)\, dy = \int_0^{\infty} \left(\int_y^{\infty} \nu(dx) \right) dy$$

$$= \int_0^{\infty} \left(\int_0^x dy \right) \nu(dx) = \int_0^{\infty} x\nu(dx)$$

$$= E(|X_1|) < \infty.$$

Hence the Borel–Cantelli lemma ensures that

$$P\{X_n = X'_n \text{ f.e.}\} = 1.$$

Therefore the conclusion of the theorem is equivalent to

$$S'_n/n \to M(\mu) \text{ a.s.}$$

This is also equivalent to

$$(*) \qquad \frac{1}{n}\left(S'_n - E(S'_n)\right) \to 0 \text{ a.s.},$$

because

$$\frac{1}{n} E(S'_n) = \frac{1}{n} \sum_{k=1}^{n} E(X'_k) = \frac{1}{n} \sum_{k=1}^{n} \int_{[-k,k]} x\mu(dx) \to M(\mu).$$

It is obvious that $X'_n \in L^2$. Hence the assertion $(*)$ follows from Theorem 4.5.1, because

$$\sum_n \frac{V(X'_n)}{n^2} \le \sum_n \frac{E\left((X'_n)^2\right)}{n^2} \le 4 \sum_n \frac{E\left((X'_n)^2\right)}{(n+1)^2}$$

$$\le 4 \sum_n \frac{1}{(n+1)^2} \int_{[0,n]} x^2 \nu(dx)$$

$$\le 4 \int_0^\infty \frac{dy}{y^2} \int_{[0,y]} x^2 \nu(dx) = 4 \int_0^\infty \left(\int_x^\infty \frac{dy}{y^2}\right) x^2 \nu(dx)$$

$$= 4 \int_0^\infty x\nu(dx) = 4 \int_{-\infty}^\infty |x|\mu(dx) < \infty. \qquad \blacksquare$$

Exercise 4.5.

    (i) Let $\varphi(t)$ be a strictly increasing positive function defined on (a, ∞) $(a > 0)$ such that

$$\int_a^\infty dt/\varphi(t)^2 < \infty.$$

    Prove that Theorem 4.5.2 holds with $V(S_n)^{1/2+\varepsilon}$ replaced by $\varphi(V(S_n))$. [*Hint*: Use the same method as in the proof of Theorem 4.5.2.]

    (ii) Prove that under the assumption of Theorem 4.5.2 we obtain a sharper result

$$\frac{S_n - E(S_n)}{\sqrt{V(S_n)}(\log V(S_n))^{1+\varepsilon}} \to 0 \text{ a.s.} \qquad (\varepsilon > 0).$$

[*Hint*: Set $\varphi(t) = \sqrt{t(\log t)^{1+\varepsilon}}$ in (i). If we set

$$\varphi(t) = \sqrt{t \log t \log^2 t \log^3 t \ldots \log^{k-1} t (\log^k t)^{1+\varepsilon}},$$

we obtain an even sharper result, where

$$\log^1 t = \log t, \qquad \log^{k+1} t = \log(\log^k t).]$$

(iii) Prove that under the assumption of Theorem 4.5.3 we obtain

$$\frac{S_n - E(S_n)}{\sqrt{n \log n \log^2 n \ldots \log^{k-1} n (\log^k n)^{1+\varepsilon}}} \to 0 \text{ a.s.}$$

(iv) Prove that

$$\lim_{n \to \infty} \int_0^1 \int_0^1 \cdots \int_0^1 \frac{x_1^q + x_2^q + \cdots + x_n^q}{x_1^p + x_2^p + \cdots + x_n^p} \, dx_1 \, dx_2 \ldots dx_n = \frac{p+1}{q+1}$$

$$(0 < p < q < \infty).$$

[*Hint*: Let $X_1, X_2, \ldots$ be independent identically distributed with uniform distribution on $[0,1]$. Then the integral is the mean value of $S_{q,n}/S_{p,n}$ $(S_{r,n} = \Sigma_1^n X_k^r, \; r = p, q)$. $n^{-1} S_{r,n} \to E(X_1^r) = (r+1)^{-1}$ a.s. by Theorem 4.5.4. Hence $S_{q,n}/S_{p,n} \to (p+1)/(q+1)$ a.s. Also, $|S_{q,n}/S_{p,n}| \le 1$. Now use the bounded convergence theorem.]

4.6 Central limit theorems

Let $A_1, A_2, \ldots$ be an independent sequence of events such that $P(A_n) = p$ $(0 < p < 1)$ for every n, and let S_n be the number of the occurring events among $A_1, A_2, \ldots, A_n$. Then the following fact is the well-known *de Moivre–Laplace theorem*:

$$P\left\{ \frac{S_n - np}{\sqrt{np(1-p)}} < s \right\} \to \int_{-\infty}^s \frac{1}{\sqrt{2\pi}} e^{-t^2/2} \, dt \qquad (n \to \infty).$$

Letting $X_n = 1_{A_n}$, we obtain an independent sequence of random variables with the identical distribution:

$$P\{X_n = 1\} = p, \qquad P\{X_n = 0\} = 1 - p.$$

The number S_n introduced above is a random variable that is expressible in terms of $\{X_n\}$ as follows:

$$S_n = \sum_{k=1}^n X_k, \qquad n = 1, 2, \ldots.$$

Since

$$E(S_n) = np, \qquad V(S_n) = np(1-p),$$

the de Moivre–Laplace theorem claims that the probability distribution of $T_n = (S_n - E(S_n))/V(S_n)^{1/2}$ converges to the Gauss distribution $N(0,1)$ as $n \to \infty$.

In the elementary theory of probability this theorem is proved by observing

$$P\{S_n = k\} = \binom{n}{k} p^k (1-p)^{n-k}$$

and using the Stirling formula: $n! \sim \sqrt{2\pi n}\, n^n e^{-n}$. Using characteristic functions, we can also prove it as follows:

$$T_n = \sum_{k=1}^{n} \frac{X_k - p}{\sqrt{npq}} \qquad (q = 1 - p),$$

$$E(e^{izT_n}) = \prod_k E\left(\exp \frac{iz(X_k - p)}{\sqrt{npq}}\right) = \left(p e^{iz\sqrt{q/np}} + q e^{-iz\sqrt{p/nq}}\right)^n$$

$$= \left\{p\left(1 + iz\sqrt{\frac{q}{np}} - \frac{qz^2}{2np}\right) + q\left(1 - iz\sqrt{\frac{p}{nq}} - \frac{pz^2}{2nq}\right) + o\left(\frac{1}{n}\right)\right\}^n$$

$$= \left(1 - \frac{z^2}{2n} + o\left(\frac{1}{n}\right)\right)^n$$

$$\to e^{-z^2/2} \qquad (n \to \infty),$$

$$(\mathcal{F}P^{T_n})(z) \to \mathcal{F}N(z) \qquad (N = N(0,1)),$$

and so

$$P^{T_n} \to N \qquad (n \to \infty)$$

by Glivenko's theorem. This proof suggests that the de Moivre–Laplace theorem may hold in a more general situation.

Let $\{X_n\} \subset L^2$ be an independent sequence of random variables and let

$$S_n = \sum_{k=1}^{n} X_k \qquad \text{and} \qquad T_n = \frac{S_n - ES_n}{\sqrt{V(S_n)}}.$$

Those theorems that give certain conditions under which the probability distribution of T_n converges to the Gauss distribution $N(0,1)$ are called *central limit theorems*. The theory of characteristic functions was developed for the purpose of proving central limit theorems. The rough idea of the proof is as follows:

$$\mathscr{F}P^{T_n}(z) = \prod_{k=1}^{n} E\left\{\exp\frac{iz(X_k - EX_k)}{\sqrt{V(S_n)}}\right\}$$

$$= \prod_{k=1}^{n}\left(1 - \frac{V(X_k)}{2V(S_n)}z^2 + \cdots\right)$$

$$\sim \prod_{k=1}^{n}\exp\left\{-\frac{V(X_k)}{2V(S_n)}z^2\right\} = \exp\left\{-\frac{z^2}{2}\right\} = \mathscr{F}N(z).$$

Let us start with some preliminary analytical facts.

Lemma 4.6.1. Suppose that for each n we are given a finite sequence of complex numbers:

$$\alpha_{n1}, \alpha_{n2}, \ldots, \alpha_{nN} \qquad (N = N(n) < \infty)$$

satisfying

$$\alpha_n \equiv \sum_k \alpha_{nk} \to \alpha, \qquad \beta_n \equiv \max_k |\alpha_{nk}| \to 0 \qquad (n \to \infty),$$

and

$$\gamma_n \equiv \sum_k |\alpha_{nk}| < b \qquad (b: \text{independent of } n).$$

Then

$$\prod_k (1 + \alpha_{nk}) \to e^\alpha \qquad (n \to \infty).$$

Proof: This is a generalization of the well-known formula

$$e^\alpha = \lim_{n \to \infty} (1 + \alpha/n)^n.$$

By the assumption $\beta_n \to 0$ we can assume that

$$|\alpha_{nk}| < \tfrac{1}{2} \qquad \text{for every } (n, k).$$

The function $\log(1 + z)$ (z: complex) is many-valued, but we take its branch vanishing at $z = 0$, which is single-valued in $|z| < 1$ and is expressed in the following power series:

$$\log(1 + z) = z - z^2/2 + z^3/3 - \cdots \qquad (|z| < 1)$$

$$= z + \theta z^2 \qquad (|z| < \tfrac{1}{2}).$$

Here θ is a complex number with absolute value ≤ 1 depending on z. To prove the lemma it suffices to observe the following:

$$\prod_k (1 + \alpha_{nk}) = \prod_k \exp\{\log(1 + \alpha_{nk})\}$$

$$= \prod_k \exp\{\alpha_{nk} + \theta_{nk}\alpha_{nk}^2\} \qquad (|\theta_{nk}| \leq 1)$$

$$= \exp\left\{\sum_k \alpha_{nk} + \sum_k \theta_{nk}\alpha_{nk}^2\right\}$$

$$= \exp\left\{\alpha_n + \beta_n\theta_n\sum_k |\alpha_{nk}|\right\} \qquad (|\theta_n| \leq 1)$$

$$= \exp\{\alpha_n + \beta_n\theta_n'b\} \qquad (|\theta_n'| \leq 1)$$

$$\to e^\alpha \qquad (n \to \infty). \qquad \blacksquare$$

Lemma 4.6.2. It holds for every $z \in \mathbb{R}$ that

$$e^{iz} = \sum_{k=1}^n \frac{(iz)^k}{k!} + \frac{\theta|z|^{n+1}}{(n+1)!} \qquad (|\theta| \leq 1).$$

Proof: Let $f(z)$ denote the left-hand side minus the first term of the right-hand side. Then we obtain

$$f^{(n+1)}(z) = i^{n+1}e^{iz}, \qquad f^{(k)}(0) = 0 \qquad (k = 0, 1, 2, \ldots, n)$$

$$f(z) = \int_0^z \int_0^{z_1} \int_o^{z_2} \cdots \int_0^{z_n} i^{n+1}e^{iz_{n+1}}\, dz_{n+1}\, dz_n \cdots dz_1,$$

and so

$$|f(z)| \leq \frac{|z|^{n+1}}{(n+1)!},$$

which completes the proof. $\blacksquare$

The following theorem may be the most useful among the several central limit theorems.

Theorem 4.6.1 (Lindeberg's central limit theorem). Let $\{X_n\} \subset L^2$ be an independent sequence with S_n its nth partial sum. If

$$\frac{1}{V(S_n)} \sum_{k=1}^n E\left((X_k - EX_k)^2, |X_k - EX_k| \geq \varepsilon\sqrt{V(S_n)}\right) \to 0 \qquad (n \to \infty)$$

for every $\varepsilon > 0$, then the probability distribution of $T_n = (S_n - E(S_n))/\sqrt{V(S_n)}$ converges to the Gauss distribution $N(0,1)$ as $n \to \infty$.

Proof: Let $Y_{nk} = (X_k - E(X_k))/\sqrt{V(S_n)}$. Then

$$E(Y_{nk}) = 0, \qquad V(Y_{nk}) = E(Y_{nk}^2), \qquad T_n = \sum_{k=1}^{n} Y_{nk},$$

$$E(T_n) = 0, \qquad \sum_{k=1}^{n} E(Y_{nk}^2) = \sum_{k=1}^{n} V(Y_{nk}) = V(T_n) = 1,$$

and the assumption of the theorem is written

$$\sum_{k=1}^{n} E(Y_{nk}^2, |Y_{nk}| \geq \varepsilon) \to 0.$$

Let μ_n denote the probability distribution of T_n. Since $\{Y_{nk}, k = 1, 2, \ldots, n\}$ is independent, we have

$$\mathcal{F}\mu_n = E(e^{iz T_n}) = \prod_k E(e^{iz Y_{nk}})$$

$$= \prod_k (1 + \alpha_{nk}) \qquad (\alpha_{nk} \equiv E(e^{iz Y_{nk}}) - 1).$$

To prove that $\mu_n \to N(0,1)$ it suffices to check that $\{\alpha_{nk}\}$ satisfies the conditions of Lemma 4.6.1 for $\alpha = -z^2/2$. Using Lemma 4.6.2, we obtain

$$\alpha_{nk} = E(e^{iz Y_{nk}}, |Y_{nk}| < \varepsilon) + E(e^{iz Y_{nk}}, |Y_{nk}| \geq \varepsilon) - 1$$

$$= E(1 + iz Y_{nk} - \tfrac{1}{2} z^2 Y_{nk}^2 + \tfrac{1}{6} \theta_3 |z Y_{nk}|^3, |Y_{nk}| < \varepsilon)$$

$$\qquad + E(1 + iz Y_{nk} + \tfrac{1}{2} \theta_2 z^2 Y_{nk}^2, |Y_{nk}| \geq \varepsilon) - 1 \qquad (|\theta_2|, |\theta_3| \leq 1)$$

$$= iz E(Y_{nk}) - \tfrac{1}{2} z^2 E(Y_{nk}^2) + \tfrac{1}{2} z^2 E(Y_{nk}^2, |Y_{nk}| \geq \varepsilon)$$

$$\qquad + \tfrac{1}{6} |z|^3 E(\theta_3 |Y_{nk}|^3, |Y_{nk}| < \varepsilon) + \tfrac{1}{2} z^2 E(\theta_2 Y_{nk}^2, |Y_{nk}| \geq \varepsilon)$$

$$= -\tfrac{1}{2} z^2 E(Y_{nk}^2) + \tfrac{1}{2} z^2 (1 + \theta_2') E(Y_{nk}^2, |Y_{nk}| \geq \varepsilon)$$

$$\qquad + \tfrac{1}{6} |z|^3 \theta_3' \varepsilon E(Y_{nk}^2) \qquad (|\theta_2'|, |\theta_3'| \leq 1).$$

But

$$\max_k E(Y_{nk}^2, |Y_{nk}| \geq \varepsilon) \leq \sum_k E(Y_{nk}^2, |Y_{nk}| \geq \varepsilon) \to 0 \qquad (n \to \infty),$$

$$\max_k E(Y_{nk}^2) \leq \varepsilon^2 + \sum_k E(Y_{nk}^2, |Y_{nk}| \geq \varepsilon) \to \varepsilon^2 \qquad (n \to \infty)$$

Hence

$$\limsup_{n \to \infty} \max_k |\alpha_{nk}| \leq \tfrac{1}{2} z^2 \varepsilon^2 + \tfrac{1}{6} |z|^3 \varepsilon^3 \to 0 \qquad (\varepsilon \downarrow 0);$$

that is,

$$\max_k |\alpha_{nk}| \to 0 \qquad (n \to \infty).$$

Also,

$$\sum_k \alpha_{nk} = -\tfrac{1}{2}z^2 + \tfrac{1}{2}z^2\big(1 + \theta_2''\big)\sum_k E\big(Y_{nk}^2, |Y_{nk}| \geq \varepsilon\big)$$

$$+ \tfrac{1}{6}|z|^3\theta_3''\varepsilon \qquad \big(|\theta_2''|, |\theta_3''| \leq 1\big).$$

Using the same argument as above, we obtain

$$\sum_k \alpha_{nk} \to -\tfrac{1}{2}z^2 \qquad (n \to \infty)$$

It is obvious that

$$\sum_k |\alpha_{nk}| \leq \tfrac{1}{2}z^2 + z^2 + \tfrac{1}{6}|z|^3\varepsilon.$$

Thus all the conditions of Lemma 4.6.1 are verified. ∎

The Lindeberg condition implies

$$v \equiv \lim_{n \to \infty} V(S_n) > 0$$

implicitly. We will prove that it also implies $v = \infty$. Suppose that $0 < v < \infty$. Then $V(X_l) > 0$ for some l. Hence

$$v_{l,\varepsilon} \equiv E\big((X_l - EX_l)^2, |X_l - EX_l| \geq \varepsilon\sqrt{V(S_n)}\big)$$

$$\geq E\big((X_l - EX_l)^2, |X_l - EX_l| \geq \varepsilon v\big) \to V(X_l) > 0 \qquad (\varepsilon \downarrow 0).$$

Therefore $v_{l,\varepsilon} > 0$ for some $\varepsilon > 0$. For this ε the left-hand side of the Lindeberg condition is no less than $v_{l,\varepsilon}/v$ and so never converges to 0 as $n \to \infty$.

Suppose that the Lindeberg condition holds. Then

$$v = \lim_{n \to \infty} V(S_n) = \infty;$$

that is,

$$\sum_{n=1}^{\infty} V(X_n) = \infty.$$

Hence Theorem 4.5.2 ensures that

$$T_{n,\alpha} \equiv \frac{S_n - E(S_n)}{V(S_n)^\alpha} \to 0 \text{ a.s.} \qquad (\alpha > \tfrac{1}{2}).$$

This implies that the probability distribution of $T_{n,\alpha}$ ($\alpha > \tfrac{1}{2}$) converges to the distribution δ. But the probability distribution of $T_{n,1/2}$ converges to $N(0,1)$ by the central limit theorem.

As we mentioned above, $T_{n,\alpha} \to 0$ a.s. for $\alpha > \frac{1}{2}$. Does $T_n = T_{n,1/2}$ converge a.s. to a random variable? Suppose it does. Then

$$T_n - T_m \to 0 \text{ a.s.} \qquad (m, n \to \infty).$$

Choose $m = m(n)$ sufficiently large so that

$$\frac{V(S_n)}{V(S_m)} \to 0 \qquad (n \to \infty)$$

and observe that

$$T_n - T_m = \frac{\sum_1^n (X_k - E(X_k))}{\sqrt{V(S_n)}} - \frac{\sum_1^m (X_k - E(X_k))}{\sqrt{V(S_m)}}$$

$$= \left(1 - \sqrt{\frac{V(S_n)}{V(S_m)}}\right) T_n - \frac{1}{\sqrt{V(S_m)}} \sum_{n+1}^m (X_k - E(X_k))$$

$$\equiv U_n - V_n.$$

As the probability distribution of T_n converges to $N(0,1)$, so does that of U_n. U_n and V_n are independent by the independence of $\{X_1, X_2,\ldots,X_m\}$. Hence

$$|E(e^{iz(T_n - T_m)})| = |E(e^{izU_n})|\,|E(e^{-izV_n})|$$

$$\le |E(e^{izU_n})| \to e^{-z^2/2} \qquad (n \to \infty).$$

Since $T_n - T_m$ ($m = m(n)$) converges to 0 a.s., the left-hand side converges to 1 as $n \to \infty$. This is a contradiction. Thus we obtain the following.

Theorem 4.6.2. Under the Lindeberg condition the sequence $T_n = (S_n - E(S_n))/\sqrt{V(S_n)}$ never converges a.s. to a random variable.

Exercise 4.6. For (i), (ii), and (iii) we assume that $\{X_n\}$ is independent, $\{X_n\} \subset L^2$, and $S_n = \sum_1^n X_k$.

   (i) Prove that the central limit theorem holds if $M \equiv \sup_n \|X_n\|_\infty < \infty$ and $\sum_n V(X_n) = \infty$. [*Hint:* $|X_n - E(X_n)| \le 2M$ a.s. Hence,

$$E\left((X_k - E(X_k))^2, |X_k - E(X_k)| \ge \varepsilon V(S_n)\right) = 0,$$

$$k = 1,2,\ldots,n,$$

holds for n sufficiently large.]

(ii) Prove that the central limit theorem holds if $\{X_n\}$ is identically distributed (μ) with $V(\mu) > 0$. [*Hint*: The left-hand side of the Lindeberg condition is

$$\frac{1}{V(\mu)} \int_{|x - M(\mu)| > \varepsilon\sqrt{nV(\mu)}} (x - M(\mu))^2 \mu(dx).]$$

(iii) Prove that the central limit theorem holds under the *Liapunov condition*:

$$\frac{1}{\sqrt{V(S_n)}^{2+\delta}} \sum_{k=1}^{n} E\left(|X_k - E(X_k)|^{2+\delta}\right) \to 0$$

$$(\delta: \text{positive constant}).$$

[*Hint*:

$$E\left(|X_k - E(X_k)|^{2+\delta}\right)$$

$$\geq E\left(|X_k - E(X_k)|^{2+\delta}, |X_k - E(X_k)| \geq \varepsilon\sqrt{V(S_n)}\right)$$

$$\geq \varepsilon^{\delta}\sqrt{V(S_n)}^{\delta} E\left(|X_k - E(X_k)|^2, |X_k - E(X_k)|\right.$$

$$\left. \geq \varepsilon\sqrt{V(S_n)}\right).]$$

(iv) Prove that

$$\lim_{n \to \infty} e^{-n} \sum_{k=0}^{n} \frac{n^k}{k!} = \frac{1}{2}.$$

[*Hint*: Consider an independent sequence of random variables $\{X_n\}$, each X_n being Poisson-distributed with mean 1. Then the central limit theorem holds by (ii). Let $S_n = \sum_{k=1}^{n} X_k$. Then $E(S_n) = n$, $V(S_n) = n$, and so the probability distribution of $T_n = (S_n - n)/\sqrt{n}$ converges to N. Since N is a continuous distribution,

$$P\{S_n \leq n\} = P\{T_n \leq 0\} \to \tfrac{1}{2}.$$

Now note that S_n is Poisson-distributed with mean n.]

4.7   The law of iterated logarithms

Let $\{X_n\} \subset L^2$ be an independent sequence and let S_n be its nth partial sum. If $V(S_n) \to \infty$, then Theorem 4.5.2 ensures that

$$\lim_{n \to \infty} \frac{S_n - E(S_n)}{V(S_n)^{1/2+\varepsilon}} = 0 \text{ a.s.} \qquad (\varepsilon > 0).$$

This shows that $|S_n - E(S_n)|$ is very small compared with $V(S_n)^{1/2+\varepsilon}$ for every $\varepsilon > 0$. However, $(S_n - E(S_n))/V(S_n)^{1/2}$ never converges a.s. to a random variable. In this respect we have the following.

Theorem 4.7.1. Under the Lindeberg condition we have

$$\limsup_{n \to \infty} \frac{S_n - E(S_n)}{V(S_n)^{1/2}} = \infty \text{ a.s.,}$$

$$\liminf_{n \to \infty} \frac{S_n - E(S_n)}{V(S_n)^{1/2}} = -\infty \text{ a.s.}$$

Proof: We can assume without any loss of generality that $E(S_n) = 0$. Since $\{X_n\}$ satisfies the Lindeberg condition, $\{X_n, X_{n+1}, \dots\}$ satisfies the same condition. Applying the central limit theorem to this sequence, we obtain

$$\lim_{m \to \infty} P\left\{\frac{S_m - S_n}{\sqrt{V(S_m - S_n)}} > \alpha\right\} = N_{0,1}(\alpha, \infty) > 0$$

$$(\alpha > 0; N_{0,1} = N(0,1))$$

for every n. Hence we can find a sequence $r(1) < r(2) < \cdots$ such that

$$\frac{\sqrt{V(S_{r(n)})}}{V(S_{r(n-1)})} \to \infty, \qquad \beta = \inf_n P\left\{\frac{S_{r(n)} - S_{r(n-1)}}{\sqrt{V(S_{r(n)} - S_{r(n-1)})}} > \alpha\right\} > 0.$$

$\{S_{r(n)} - S_{r(n-1)}, \ n = 1, 2, \dots\}$ is obviously independent. Hence Borel's theorem ensures that

$$S_{r(n)} - S_{r(n-1)} > \alpha\sqrt{V(S_{r(n)} - S_{r(n-1)})} \text{ (i.o.) a.s.}$$

Hence

$$\frac{S_{r(n)}}{\sqrt{V(S_{r(n)})}} > \frac{V(S_{r(n-1)})}{\sqrt{V(S_{r(n)})}}\frac{S_{r(n-1)}}{V(S_{r(n-1)})} + \alpha\sqrt{\frac{V(S_{r(n)} - S_{r(n-1)})}{V(S_{r(n)})}} \text{ (i.o.) a.s.}$$

Let $n \to \infty$. Both factors of the first term of the right-hand side converge to 0 by the choice of $\{r(n)\}$ and Theorem 4.5.2 ($\varepsilon = \frac{1}{2}$), and the second term converges to α because

$$V\left(S_{r(n)} - S_{r(n-1)}\right) = V\left(S_{r(n)}\right) - V\left(S_{r(n-1)}\right).$$

Hence

$$\limsup_{n \to \infty} \frac{S_{r(n)}}{\sqrt{V\left(S_{r(n)}\right)}} \geq \alpha \text{ a.s.,}$$

and so

$$\limsup_{n \to \infty} \frac{S_n}{\sqrt{V\left(S_n\right)}} \geq \alpha \text{ a.s.}$$

Letting $\alpha \uparrow \infty$, we obtain the first assertion of the theorem. The second assertion follows from this by replacing X_n by $-X_n$. ∎

Due to this theorem, $S_n - E(S_n)$ exceeds the levels $\sqrt{V(S_n)}$ and $-\sqrt{V(S_n)}$ infinitely often almost surely. From this fact we can see that if

$$P\{X_n = -1\} = P\{X_n = 1\} = 1/2,$$

then $S_n = 0$ infinitely often almost surely.

Under a stronger assumption than the Lindeberg condition we obtain the following estimates:

$$\limsup_{n \to \infty} \frac{S_n - E(S_n)}{\left(2V(S_n)\log^2 V(S_n)\right)^{1/2}} = 1 \text{ a.s.,}$$

$$\liminf_{n \to \infty} \frac{S_n - E(S_n)}{\left(2V(S_n)\log^2 V(S_n)\right)^{1/2}} = -1 \text{ a.s.,}$$

$$\limsup_{n \to \infty} \frac{|S_n - E(S_n)|}{\left(2V(S_n)\log^2 V(S_n)\right)^{1/2}} = 1 \text{ a.s.}$$

This is a much more precise result than the theorem above and is called the *law of iterated logarithms*. This astonishing fact was first proved by A. Khinchin and was made more precise later. Here we discuss the case where every X_n is a *Gaussian variable* (a Gauss-distributed random variable) to clarify the idea of the proof.

Theorem 4.7.2. Suppose that $\{X_n\}$ is an independent sequence of Gaussian variables. If $V(S_n) \to \infty$ and $V(X_n)/V(S_n) \to 0$, then the law of iterated logarithm holds.

Proof: We can assume that $E(X_n) = 0$ (and so $E(S_n) = 0$)). First we will prove

(i) $$\limsup_{n \to \infty} \frac{S_n}{\left(2V(S_n)\log^2 V(S_n)\right)^{1/2}} \leq 1 \text{ a.s.}$$

Define a sequence of natural numbers $n(1) < n(2) < n(3) < \cdots$ for $\varepsilon > 0$ as follows:

$$n(1) = 1, \qquad n(k) = \min\left\{ n \,|\, V(S_n) \geq (1+\varepsilon)V\left(S_{n(k-1)}\right) \right\}$$

Then

$$V\left(S_{n(k)}\right) \geq (1+\varepsilon)V\left(S_{n(k-1)}\right) > V\left(S_{n(k)-1}\right).$$

The assumption $V(X_n)/V(S_n) \to 0$ implies $V(S_n)/V(S_{n-1}) \to 1$. Hence, for k sufficiently large, say $k > k_0 = k_0(\varepsilon)$, we have

$$V\left(S_{n(k)}\right) < (1+\varepsilon)V\left(S_{n(k)-1}\right) < (1+\varepsilon)^2 V\left(S_{n(k-1)}\right).$$

Then we can find $c_1 = c_1(\varepsilon) > 0$ such that

$$c_1(1+\varepsilon)^k < V\left(S_{n(k)}\right) < c_1(1+\varepsilon)^{2k},$$

where c_1 does not depend on k. (Hereafter, $c_2, c_3, \ldots$ denote positive constants independent of k.)

Letting

$$T_k = \max\left\{ S_j \,|\, 1 \leq j \leq n(k) \right\}, \qquad k > k_0,$$

we estimate the probabilities

$$p_k = p_k(\varepsilon, \delta) = P\left\{ T_k \geq \left(2(1+\delta)V\left(S_{n(k)}\right)\log^2 V\left(S_{n(k)}\right)\right)^{1/2} \right\} \qquad (\delta > 0).$$

By the assumption it is obvious that $S_n - S_j$ is Gauss-distributed with mean 0. Hence

$$P\{ S_n - S_j \geq 0 \} \geq \tfrac{1}{2} \qquad (n \geq j).$$

For every $a > 0$ we have

$$P\left\{ \max_{j=1}^{n} S_j \geq a \right\} = \sum_{k=1}^{n} P\left\{ \max_{j=1}^{k-1} S_j < a \leq S_k \right\} \qquad \left(\max_{j=1}^{k-1} S_j = 0 \text{ for } k = 1 \right),$$

$$P\{ S_n \geq a \} \geq \sum_{k=1}^{n} P\left\{ \max_{j=1}^{k-1} S_j < a \leq S_k, S_n - S_k \geq 0 \right\}$$

$$= \sum_{k=1}^{n} P\left\{ \max_{j=1}^{k-1} S_j < a \leq S_k \right\} P\{ S_n - S_k \geq 0 \}$$

(by independence of $\{ X_k \}$)

$$\geq \tfrac{1}{2} \sum_{k=1}^{n} P\left\{ \max_{j=1}^{k-1} S_j < a \leq S_k \right\}.$$

Hence

$$P\left\{\max_{j=1}^{n} S_j \geq a\right\} \leq 2P\{S_n \geq a\}.$$

Replacing a by $a\sqrt{V(S_n)}$, we obtain

$$P\left\{\max_{j=1}^{n} S_j \geq a\sqrt{V(S_n)}\right\} \leq 2P\left\{S_n \geq a\sqrt{V(S_n)}\right\} = 2\int_a^\infty \frac{1}{\sqrt{2\pi}} e^{-x^2/2} dx,$$

$$\int_a^\infty e^{-x^2/2} dx = \int_a^\infty \frac{1}{x} e^{-x^2/2} x \cdot dx = -\frac{1}{x} e^{-x^2/2}\Big|_a^\infty - \int_a^\infty \frac{1}{x^2} e^{-x^2/2} dx$$

$$\leq \frac{1}{a} e^{-a^2/2} \leq e^{-a^2/2} \qquad (a > 1),$$

$$P\left\{\max_{j=1}^{n} S_j \geq a\sqrt{V(S_n)}\right\} \leq e^{-a^2/2} \qquad (a > 1).$$

Setting $n = n(k)$ and $a = (2(1 + \delta)\log^2 V(S_{n(k)}))^{1/2}$ in this inequality, we obtain

$$p_k \equiv P\left\{T_k \geq \left(2(1 + \delta)V(S_{n(k)})\log^2 V(S_{n(k)})\right)^{1/2}\right\}$$

$$\leq \exp\left\{-(1 + \delta)\log^2 V(S_{n(k)})\right\}$$

$$= \left(\log V(S_{n(k)})\right)^{-(1+\delta)}$$

$$\leq \left(\log\left(c_1(1 + \varepsilon)^k\right)\right)^{-(1+\delta)}$$

$$\leq c_2 k^{-(1+\delta)},$$

and so

$$\sum_k p_k < \infty.$$

Hence the Borel–Cantelli lemma ensures that

$$P\left\{T_k < \left(2(1 + \delta)V(S_{n(k)})\log^2 V(S_{n(k)})\right)^{1/2} \text{ f.e.}\right\} = 1;$$

in other words,

(ii) $$\limsup_{k \to \infty} \frac{T_k}{\left(2V(S_{n(k)})\log^2 V(S_{n(k)})\right)^{1/2}} \leq (1 + \delta)^{1/2} \text{ a.s.}$$

For every $n \in [n(k-1), n(k))$ we obtain

$$\frac{S_n}{\left(2V(S_n)\log^2 V(S_n)\right)^{1/2}} \leq \frac{T_k}{\left(2V(S_{n(k-1)})\log^2 V(S_{n(k-1)})\right)^{1/2}}$$

$$= \frac{T_k}{\left(2V(S_{n(k)})\log^2 V(S_{n(k)})\right)^{1/2}}$$

$$\times \left[\frac{V(S_{n(k)})}{V(S_{n(k-1)})} \cdot \frac{\log^2 V(S_{n(k)})}{\log^2 V(S_{n(k-1)})}\right]^{1/2}.$$

Since $\log^2 x/x$ is decreasing in $x > e^e$,

$$e^e < x < y \Rightarrow \frac{\log^2 y}{y} < \frac{\log^2 x}{x} \Rightarrow \frac{\log^2 y}{\log^2 x} < \frac{y}{x}.$$

Hence the factor $[\ ]^{1/2}$ is no more than

$$\frac{V(S_{n(k)})}{V(S_{n(k-1)})} \qquad \left(< (1+\varepsilon)^2\right)$$

whenever $n(k)$ is sufficiently large. By virtue of (ii) we obtain

$$\limsup_{n\to\infty} \frac{S_n}{\left(2V(S_n)\log^2 V(S_n)\right)^{1/2}} \leq (1+\delta)^{1/2}(1+\varepsilon)^2 \text{ a.s.}$$

Letting $\varepsilon \downarrow 0$ and $\delta \downarrow 0$, we obtain (i).

Replacing $\{X_n\}$ by $\{-X_n\}$ in (i), we obtain

$$\liminf_{n\to\infty} \frac{S_n}{\left(2V(S_n)\log^2 V(S_n)\right)^{1/2}} \geq -1 \text{ a.s.},$$

which, combined with (i), implies that

(iii) $$\limsup_{n\to\infty} \frac{|S_n|}{\left(2V(S_n)\log^2 V(S_n)\right)^{1/2}} \leq 1 \text{ a.s.}$$

Next we will prove that

(iv) $$\limsup_{n\to\infty} \frac{S_n}{\left(2V(S_n)\log^2 V(S_n)\right)^{1/2}} \geq 1 \text{ a.s.}$$

This, combined with (i), implies the first formula of the theorem. The second formula follows by replacing $\{X_n\}$ by $\{-X_n\}$, and these two formulas imply the third one.

To prove (iv) we take for every $\alpha > 1$ a sequence of natural numbers $1 = m(1) < m(2) < m(3) < \cdots$ such that

(v) $\qquad V(S_{m(k)}) \geq \alpha V(S_{m(k-1)}) > V(S_{m(k)-1})$

Using the same argument as for the sequence $\{n(k)\}$ above, we obtain

(vi) $\qquad c_2 \alpha^{2k} > V(S_{m(k)}) > c_3 \alpha^k$.

As $\{X_k\}$ is an independent sequence of Gaussian variables, so is the following sequence:

$$\Delta_k = S_{m(k)} - S_{m(k-1)}, \qquad k = 1, 2, \ldots.$$

Now we will estimate

$$q_k = q_k(\alpha, \delta) = P\{\Delta_k > (2(1-\delta)V(\Delta_k)\log^2 V(\Delta_k))^{1/2}\}$$

for $\delta \in (0, 1)$. If a is sufficiently large, we obtain

$$P\{\Delta_k > a(V(\Delta_k))^{1/2}\} = \int_a^\infty \frac{1}{\sqrt{2\pi}} e^{-x^2/2} \, dx,$$

$$= \frac{1}{\sqrt{2\pi}} \left(\frac{1}{a} e^{-a^2/2} - \frac{1}{a^3} e^{-a^2/2} + \int_a^\infty \frac{3}{x^4} e^{-x^2/2} \, dx \right)$$

(by applying integration by parts twice)

$$\geq \frac{1}{4a} e^{-a^2/2}.$$

For k sufficiently large we set $a = (2(1-\delta)\log^2 V(\Delta_k))^{1/2}$ to obtain

$$q_k \geq c_4 (\log^2 V(\Delta_k))^{-1/2} (\log V(\Delta_k))^{-1+\delta}$$

$$\geq c_5 (\log V(\Delta_k))^{-1} \geq c_5 (\log V(S_k))^{-1}$$

because

$$\lim_{x \to \infty} (\log^2 x)^{-1/2} (\log x)^\delta = \infty.$$

Using (vi) we obtain $q_k \geq c_6 k^{-1}$ for k sufficiently large. This implies that $\Sigma_k q_k = \infty$. Since $\{\Delta_k\}$ is independent, Borel's theorem ensures that

$$P\{\Delta_k > (2(1-\delta)V(\Delta_k)\log^2 V(\Delta_k))^{1/2} \text{ i.o.}\} = 1,$$

and so

(vii) $\qquad \displaystyle\limsup_{n \to \infty} \frac{\Delta_k}{(2V(\Delta_k)\log^2 V(\Delta_k))^{1/2}} \geq (1-\delta)^{1/2}$ a.s.

But

$$\frac{S_{m(k)}}{\left(2V(S_{m(k)})\log^2 V(S_{m(k)})\right)^{1/2}} \geq \frac{\Delta_k}{\left(2V(S_{m(k)})\log^2 V(S_{m(k)})\right)^{1/2}}$$
$$- \frac{|S_{m(k-1)}|}{\left(2V(S_{m(k)})\log^2 V(S_{m(k)})\right)^{1/2}},$$

$$V(\Delta_k), V(S_{m(k-1)}) \leq V(S_{m(k)}),$$

$$\frac{V(\Delta_k)}{V(S_{m(k)})} \geq 1 - \frac{1}{\alpha}, \qquad \frac{V(S_{m(k-1)})}{V(S_{m(k)})} \leq \frac{1}{\alpha},$$

and

$$e^e < x < y \Rightarrow 1 < \frac{\log^2 y}{\log^2 x} < \frac{y}{x}.$$

Hence

$$\frac{S_{m(k)}}{\left(2V(S_{m(k)})\log^2 V(S_{m(k)})\right)^{1/2}} \geq \frac{\Delta_k}{\left(2V(\Delta_k)\log^2 V(\Delta_k)\right)^{1/2}}\left(1 - \frac{1}{\alpha}\right)$$

$$- \frac{|S_{m(k-1)}|}{\left(2V(S_{m(k-1)})\log^2 V(S_{m(k-1)})\right)^{1/2}}\frac{1}{\alpha},$$

and so

$$\limsup_{k \to \infty} \frac{S_{m(k)}}{\left(2V(S_{m(k)})\log^2 V(S_{m(k)})\right)^{1/2}} \geq (1-\delta)^{1/2}\left(1 - \frac{1}{\alpha}\right) - \frac{1}{\alpha} \text{ a.s.}$$

by (iii) and (vii). Therefore,

$$\limsup_{n \to \infty} \frac{S_n}{\left(2V(S_n)\log^2 V(S_n)\right)^{1/2}} \geq (1-\delta)^{1/2}\left(1 - \frac{1}{\alpha}\right) - \frac{1}{\alpha} \text{ a.s.}$$

Letting $\delta \downarrow 0$ and $\alpha \uparrow \infty$ we obtain (iv). ∎

Exercise 4.7. The first formula of Theorem 4.7.2 shows that

$$P\left\{S_n - E(S_n) < (1+\delta)\left(2V(S_n)\log^2 V(S_n)\right)^{1/2} \text{ f.e.}\right\} = 1,$$

$$P\left\{S_n - E(S_n) > (1-\delta)\left(2V(S_n)\log^2 V(S_n)\right)^{1/2} \text{ i.o.}\right\} = 1$$

for every $\delta > 0$. The key of the proof was to prove

$$\sum_k p_k < \infty, \qquad \sum_k q_k = \infty,$$

using the fact that

$$\frac{1}{4a}e^{-a^2/2} \leq P(U > a\sqrt{V}) \leq e^{-a^2/2}$$

for a Gaussian variable U with mean 0 and variance V. Use the same technique to obtain the following more precise results:

$$P\left\{S_n - E(S_n) < \left(2V(S_n)\left(\log^2 V(S_n) + (1+\delta)\log^3 V(S_n)\right)\right)^{1/2} \text{ f.e.}\right\} = 1,$$

$$P\left\{S_n - E(S_n) > \left(2V(S_n)\left(\log^2 V(S_n) + (1-\delta)\log^3 V(S_n)\right)\right)^{1/2} \text{ i.o.}\right\} = 1$$

for every $\delta > 0$ and similarly for the results corresponding to the second and the third formulas.

4.8 Gauss's theory of errors

Whenever we measure a quantity, there occur some errors. There are two kinds of errors, systematic ones and random ones. Systematic errors can be avoided by exercising care. For example, let us consider the error caused by slight difference in the lengths of two arms of a scale when weighing an object. Let d_1 and d_2 denote the lengths of the arms. Suppose that we obtained the weight w_1 when putting the object on one side and w_2 when putting it on the other side. If the true weight is w, then

$$w_1 = wd_1/d_2, \qquad w_2 = wd_2/d_1$$

and so

$$w = \sqrt{w_1 w_2}.$$

Thus the geometric mean of two observed values gives the true value.

On the other hand, a random error is the sum of a large number of small uncontrollable errors due to the molecular motion of the air, the slight vibration of the floor, the delicate change of gravity, and so on. Gauss's theory of errors started as a method to remove random errors, and is based on the assumption that a random error is regarded as a Gaussian random variable with mean 0.

To justify this assumption, it may be most natural to regard a random error to be the sum of many small independent random errors and to prove that it is nearly Gauss-distributed due to the following theorem.

Theorem 4.8. Suppose that for every n, $X_{n1}, X_{n2}, \cdots, X_{nN}$ $(N = N(n))$ is independent, and $S_n = \sum_{k=1}^{N(n)} X_{nk}$. If

$$\varepsilon_n = \sup_{k,\omega} |X_{nk}(\omega)| \to 0,$$

$$E(S_n) \equiv \sum_k E(X_{nk}) \to m,$$

$$V(S_n) \equiv \sum_k V(X_{nk}) \to v,$$

then the probability distribution μ_n of S_n converges to $N(m, v)$ as $n \to \infty$.

Proof: $\theta_{nk}, \theta'_{nk}, \ldots$ represent complex numbers with absolute values ≤ 1. By independence of $\{X_{nk}, k = 1, 2, \ldots, N\}$ we can use the same argument as in the proof of the central limit theorem to obtain

$$\mathcal{F}\mu_n(z) = E(e^{izS_n}) = \prod_k E(e^{izX_{nk}}).$$

$$= e^{izm_n}\prod_k E(e^{iz(X_{nk}-m_{nk})})$$

$$\left(m_{nk} = E(X_{nk}), m_n = E(S_{nk}), m_n = \sum_k m_{nk}\right)$$

$$= e^{izm_n}\prod_k\left(1 - \frac{z^2}{2}V(X_{nk}) + \frac{z^3}{6}E(|X_{nk} - m_{nk}|^3)\theta_{nk}\right)$$

$$= e^{izm_n}\prod_k\left(1 - \frac{z^2}{2}V(X_{nk}) + \frac{z^3}{6}V(X_{nk})(2\varepsilon_n)\theta'_{nk}\right)$$

$$\left(\text{because } |X_{nk} - m_{nk}|^3 \leq (X_{nk} - m_{nk})^2(|X_{nk}| + |m_{nk}|)\right)$$

$$\rightarrow \exp\left(izm - \frac{z^2}{2}v\right). \quad\blacksquare$$

Thus the probability distribution of the random errors is of the form $N(m, v)$. But m may be assumed to vanish; otherwise there must be a systematic error that should have been removed in advance.

Let X be a value observed when measuring a quantity whose true value is a. Then $X - a$ is the random error that is $N(0, v)$-distributed. In view of the fact that

$$P\{|X - a| > t\sqrt{2v}\} = 2\int_{t\sqrt{2}}^{\infty}\frac{1}{\sqrt{2\pi}}e^{-s^2/2}\,ds = \frac{2}{\sqrt{\pi}}\int_t^{\infty}e^{-s^2}\,ds$$

is determined only by t, $h = 1/\sqrt{2v}$ is called the *accuracy* of the measurement; in fact, the larger h is, the more accurate the measurement is.

In order to raise the accuracy, one usually repeats the measurement a number of times and takes the average of the observed values. This procedure is justified as follows.

Let $X_1, X_2, \ldots, X_n$ be the values observed. Then these random variables are independently identically distributed $(N(a, v))$. Hence the mean values of $X_1, X_2, \ldots, X_n$, written $\bar{X}$, is given by

$$\bar{X} = \frac{1}{n}(X_1 + X_2 + \cdots + X_n).$$

Note that $\overline{X}$ is also a random variable. Since

$$E(e^{iz\overline{X}}) = \prod_{k=1}^{n} E(e^{izX_k/n})$$

$$= \prod_{k=1}^{n} \exp\left\{\frac{iaz}{n} - \frac{vz^2}{2n^2}\right\}$$

$$= \exp\left\{iaz - \frac{v}{2n}z^2\right\},$$

$\overline{X}$ is $N(a, v/n)$-distributed. Hence $\overline{X}$ is equivalent to a measurement whose accuracy is $\sqrt{n}/\sqrt{2v}$ (i.e., $\sqrt{n}$ times the accuracy of each measurement).

This fact results from the assumption that the random variables are Gauss-distributed with mean a. If they are Cauchy-distributed with central value a, then

$$P\{X_k \in E\} = \frac{c}{\pi}\int_E \frac{dx}{c^2 + (x - a)^2}, \qquad k = 1, 2, \ldots, n,$$

$$E(e^{izX_k}) = e^{iaz - c|z|},$$

$$E(e^{iz\overline{X}}) = \left(\exp\left\{\frac{iaz}{n} - \frac{c|z|}{n}\right\}\right)^n = e^{iaz - c|z|},$$

$$P^{\overline{X}} = P^{X_k}, \qquad k = 1, 2, \ldots, n.$$

Hence the average of many observed values is equivalent to one observed value. Though the graph of Gauss density appears similar to that of Cauchy density, there is an essential difference in this respect.

Returning to the Gaussian case, we consider the weighted average of a number of observed values:

$$\overline{X}_\alpha = \sum_k \alpha_k X_k,$$

where

$$(*) \qquad \alpha_1, \alpha_2, \ldots, \alpha_n \geqq 0, \qquad \sum_{k=1}^{n} \alpha_k = 1.$$

Then $\overline{X}$ is $N(m, v_\alpha)$ distributed, where $v_\alpha = \sum_k \alpha_k^2 v$. Since v_α turns out to be smallest when $\alpha_1 = \alpha_2 = \cdots = \alpha_n = 1/n$, the average with equal weight is the optimum estimate.

Exercise 4.8. Suppose that we obtained X_1, X_2, X_3 when measuring the inner angles of a triangle, $\theta_1, \theta_2, \theta_3$, in the same way independently. Find the optimum linear expressions of X_1, X_2, X_3 to estimate $\theta_1, \theta_2, \theta_3$, respectively.

[*Hint*: The random errors $X_1 - \theta_1$, $X_2 - \theta_2$, $X_3 - \theta_3$ are independently identically distributed ($N(0, v)$). Let

$$\bar{X}_i = a_i + \sum_{k=1}^{3} a_{ik} X_k, \qquad i = 1, 2, 3,$$

be the optimum ones. Then $\bar{X}_1$ is $N(\bar{\theta}_1, v_1)$ – distributed, where

$$\bar{\theta}_1 = a_1 + \sum_k a_{1k} \theta_1, \qquad v_1 = \sum_k a_{1k}^2 v.$$

Since θ_1 is the true value, we shall have $\bar{\theta}_1 = \theta_1$ (i.e., $a_1 + \sum_k a_{1k} \theta_k = \theta_1$). This must hold whenever $\theta_1 + \theta_2 + \theta_3 = \pi$. Hence

$$\frac{a_{11} - 1}{1} = \frac{a_{12}}{1} = \frac{a_{13}}{1} = \frac{-a_1}{\pi}.$$

Under this condition we have $v_1/v = \sum_k a_{1k}^2 = a_{11}^2 + 2(a_{11} - 1)^2 = 3a_{11}^2 - 4a_{11} + 2$, which is minimum if $a_{11} = \frac{2}{3}$. Hence, we obtain $a_{12} = -\frac{1}{3}$, $a_{13} = -\frac{1}{3}$, and $a_1 = \pi/3$. Hence the optimum estimate of θ_1 is given by

$$\bar{X}_1 = \pi/3 + \tfrac{2}{3} X_1 - \tfrac{1}{3} X_2 - \tfrac{1}{3} X_3.$$

Similarly

$$\bar{X}_2 = \pi/3 - \tfrac{1}{3} X_1 + \tfrac{2}{3} X_2 - \tfrac{1}{3} X_3,$$

$$\bar{X}_3 = \pi/3 - \tfrac{1}{3} X_1 - \tfrac{1}{3} X_2 + \tfrac{2}{3} X_3.$$

Note that $\bar{X}_1 + \bar{X}_2 + \bar{X}_3 = \pi$ and $V(\bar{X}_i) = \frac{2}{3} v < v = V(X_i)$.]

4.9 Poisson's law of rare events

Suppose that $\alpha_1, \alpha_2, \ldots, \alpha_n$ are independent events, each α_i occurring with small probability p_k, where $\sum_{k=1}^{n} p_k$ is nearly a constant λ. Poisson's law of rare events claims that the probability distribution of the number of occurrences among these events is nearly Poisson-distributed with mean λ. For example, if (1) an accident occurs with a very small probability in a factory on each day and (2) if the occurrence on a particular day does not influence that on another day, then the number of accidents for one year is nearly Poisson-distributed.

Let X_k be a random variable taking 0 or 1 according to whether α_k occurs or not. Then $X_1, X_2, \ldots, X_n$ are independent by independence of $\alpha_1, \alpha_2, \ldots, \alpha_n$ and the number of occurrences of $\{\alpha_k\}$ is given by $N = X_1 + X_2 + \cdots + X_n$. Hence the mathematical formulation of Poisson's law is as follows.

Theorem 4.9 (Poisson's law of rare events). Suppose that $X_{n1}, X_{n2}, \ldots, X_{nm}$ $(m = m(n) \to \infty)$ are independent for each n and

$$P(X_{nk} = 1) = p_{nk} \quad \text{and} \quad P(X_{nk} = 0) = 1 - p_{nk}.$$

If $\bar{p}_n \equiv \max_k p_{nk} \to 0$ and $p_n \equiv \Sigma_k p_{nk} \to \lambda$ $(n \to \infty)$, then the probability distribution of $N_n = \Sigma_{k=1}^m X_{nk}$ converges to the Poisson distribution p_λ.

Proof: By the assumption we have

$$E(e^{izN_n}) = \prod_k E(e^{izX_{nk}}) = \prod_k \left(p_{nk}e^{iz} + (1 - p_{nk}) \right)$$

$$= \prod_k \left(1 + p_{nk}(e^{iz} - 1) \right)$$

$$= \prod_n \left(1 + \alpha_{nk} \right).$$

But

$$\max_k |\alpha_{nk}| \leq 2 \max_k p_{nk} \to 0,$$

$$\sum_k |\alpha_{nk}| \leq 2 \sum_k p_{nk} \to 2\lambda,$$

$$\sum_k \alpha_{nk} = \sum_k p_{nk}(e^{iz} - 1) \to \lambda(e^{iz} - 1).$$

Hence we can use Lemma 4.6.1 to conclude

$$E(e^{izN_n}) \to \exp\{\lambda(e^{iz} - 1)\} = \mathscr{F}p_\lambda(z),$$

which completes the proof. ∎

Exercise 4.9.

(i) Prove the special case ($p_{nk} = \lambda/m$, $m = m(n)$) of Poisson's law of rare events in an elementary way. [*Hint*: Note that

$$P\{N_n = k\} = \binom{m}{k} \left(\frac{\lambda}{m}\right)^k \left(1 - \frac{\lambda}{m}\right)^{m-k}$$

$$= \frac{m(m-1)\ldots(m-k+1)}{m^k} \left(1 - \frac{\lambda}{m}\right)^m \left(1 - \frac{\lambda}{m}\right)^{-k} \frac{\lambda^k}{k!}$$

$$\to e^{-\lambda}\lambda^k/k! \quad (k \text{ fixed}, m \to \infty).]$$

(ii) Suppose that $X_1, X_2, \ldots, X_k$ are independent, and each X_k is Poisson(p_λ)-distributed. Then the conditional probability distribution of $(X_1, X_2,\ldots,X_k)$ under $X_1 + X_2 + \cdots + X_k = n$ is a multinomial distribution:

$$\left(\frac{n!}{n_1!n_2!\ldots n_k!} \prod_{i=1}^{k} \left(\frac{\lambda_i}{\lambda_1 + \lambda_2 + \cdots + \lambda_n} \right)^{n_i} \right) \quad \left(\sum_{i=1}^{n} n_i = n \right).$$

[*Hint*:

$$P\left\{ (X_1, X_2,\ldots,X_k) = (n_1, n_2,\ldots,n_k) \Big| \sum_i X_i = n \right\} \quad \left(n = \sum n_i \right)$$

$$= \frac{P\{ (X_1, X_2,\ldots,X_k) = (n_1, n_2,\ldots,n_k) \}}{P\left\{ \sum_i X_i = n \right\}}$$

$$= \frac{\prod_i P\{ X_i = n_i \}}{P\left\{ \sum_i X_i = n \right\}}.$$

Since $\sum_i X_i$ is Poisson-distributed with parameter $\lambda \equiv \sum \lambda_i$, this ratio is equal to

$$\frac{\prod_i e^{-\lambda_i} \lambda_i^{n_i}/n_i!}{e^{-\lambda} \lambda^n/n!} = \frac{n!}{n_1!n_2!\ldots n_k!} \prod_i \left(\frac{\lambda_i}{\lambda} \right)^{n_i}.]$$

Index

absolute moments: of distributions, 76; of random variables, 154
accuracy, 207

Bayes's theorem, 27
Bienaymé's inequality: on distributions, 77; on random variables, 156
Bochner's theorem, 91
Borel–Cantelli lemma, 115
Borel-measurable maps, 45
Borel sets, 43
Borel systems, 43
Borel's theorem, 134

Carathéodory's theorem, 48
central limit theorem, 192; Liapunov's, 198; Lindeberg's, 194
central values: of distributions, 183; of random varibles, 176; *see also* de Moivre–Laplace theorem
characteristic functions: of distributions, 80; of random variables, 154, 155
Chebyshev's inequality, 12, 154
coincidence theorem of probability measures, 47
component variables, 5
conditional distributions, 147
conditional expectations, *see* conditional means
conditional mean operators, 157
conditional means, 147
conditional probability measures, 24; under decompositions, 138; under random variables, 143; under σ-algebras, 143
consistency condition, Kolmogorov's, 106
convergence of random variables: almost sure, 149; in distributions, 150; in mean, 150; in probability, 148
convolution, 77
coordinate representations, 165
correlation coefficients, 10, 154
covariances, 9, 154

decompositions, 123–4; random variables and, 125; separable, 123; sequence of, 124–6; σ-algebras and, 126–8; *see also* conditional probability measures
de Moivre–Laplace theorem, 191
dispersions: of distributions, 183; of random variables, 176
distribution functions: of distributions, 52, 104; of random variables, 154
distributions, 67, 102, 105; absolutely continuous, 68; continuous, 68; convergence of, 70; *d*-dimensional, 102; infinite-dimensional, 106; Lebesgue decomposition of, 69; purely discontinuous, 67; reflection of, 81; *see also* characteristic functions, convolution, distribution functions, weak topology in the distributions
Dynkin class theorem, 46

elementary probability measures, 49
events, 1, 114; almost sure occurrence of, 10; complementary, 1; difference, 1; direct sum, 2; exclusive, 2; intersection, 1; occurrence of, 1; proper difference, 1; sum, 1; *see also* independence, probability
expectation, *see* mean values
extension theorem of probability measures, 50; *see also* Kolmogorov's extension theorem

formula on change of variables, 122
Fourier transforms: of distributions, 80; of functions, 94

Gauss's theory of errors, 206
generating functions, 34; multiplicativity of, 34
Glivenko's theorem, 87

Hölder's inequality, 76, 159

inclusion–exclusion formulas, 3
inclusion–maps, 111
independence, 27, 129; of composition
    of trials, 32; of events, 30; of random
    variables, 27, 131; of σ-algebras, 129
indicators, 8
integration by parts, generalization of, 85

Jensen's inequality, 160
joint variables, 5, 120; topological, 120

Kac's theorem, 155
Khinchin's three series theorems, 171
Kolmogorov's extension theorem, 106
Kolmogorov's inequality, 167
Kolmogorov's theorem on convergent series
    of independent random variables, 170
Kronecker's lemma, 187

law of iterated logarithms, 200
law of large numbers, 35–6; Bernoulli's,
    37; Kolmogorov's strong, 188; *see
    also* strong law of large numbers for
    independently identically distributed
    random variables
Lebesgue–Stieltjes measures, 51
Lévy metric, 100
Lévy's convergence theorem, 90
Lévy's equivalence theorem, 173
Luzin's theorem, 63

mean values, 121; additivity of, 7;
    multiplicativity of, 32; properties of,
    121
mean vectors, 155
measures, 39
moments: of distribution, 76; of random
    variables, 154
multiplicative classes, 46

null sets, 45

Ottaviani's inequality, 168
outer measures, 47

Poisson's law of rare events, 210
Polish spaces, 100
Polya's theorem, 94
positive-definite functions, 85
principle of maximum entropy, 17
probability, 2
probability density, 39
probability measures, 2, 39; Borel, 47;
    complete, 45; complete direct product
    of, 59; direct product of, 53; K-regular,
    61; Lebesgue extension of, 45; perfect,
    110; regular, 47; separable, 111;

standard, 61; *see also* distributions,
    probability spaces
probability spaces, 2, 39; complete, 45;
    isomorphic, 60; perfect, 110; separable,
    111; standard, 61; strictly isomorphic,
    60
product maps, 45
projection, 6

random variables, 13, 116; extended real,
    120; joint probability distributions
    of, 5, 116; probability distributions
    of, 4, 116; probability spaces of, 4,
    119; P-separable, 117; real, 3, 16;
    topological, 117; vector-valued, 5,
    116; *see also* component variables,
    correlation coefficients, covariances,
    decompositions, independence, joint
    variables, mean values, moments,
    variances

sample points, 1
sample spaces, 1
sampling, 39
separating classes, 110
sequence spaces, 65
series of independent random variables,
    165; of convergent type, 184; of
    properly divergent type, 184; of
    pseudodivergent type, 184
σ-algebras, 38; product, 43; σ-generated,
    113; *see also* condition probability
    measures, decompositions,
    independence
special distributions: bionomial, 68; Cantor,
    69; Cauchy, 41; δ, 68; exponential,
    79; Gauss, 41, 103, 105; multinomial,
    104; Poisson, 68; triangular, 79;
    uniform, 79
standard deviations, 10, 154
strong law of large numbers for
    independently identically distributed
    random variables, 189

trials, 1; direct decomposition, 15; mixing
    of, 14; multiplicative law, 16, 22;
    probability spaces of, 1; tree
    composition of, 18

uniform integrability, 151

variances, 9, 154; additivity of, 33, 154

weak topology in the distributions, 98

zero-one law, Kolmogorov's, 131